A Beekeeper's Life
Tales from the Bottom Board

Ed Colby

NB

Northern Bee Books

A Beekeeper's Life - Tales from the Bottom Board
© Ed Colby 2021

Published in the United Kingdom by

Northern Bee Books,

Scout Bottom Farm,

Mytholmroyd,

West Yorkshire HX7 5JS

Tel: 01422 882751

Fax: 01422 886157

www.northernbeebooks.co.uk

ISBN 978-1-912271-88-7

Cover photograph © John Kelly
Back cover photograph courtesy of Marilyn Gleason

Photographs with this book are courtesy of Irene Owsley, John Kelly, Marilyn Gleason, Tracy Delli Quadri and the author.

Design and artwork, DM Design and Print

A Beekeeper's Life
Tales from the Bottom Board

To friend and mentor Paul Limbach, who led me down this crooked road,
and gave me hope, always.

Contents

Testimonials

"Ed's column is the last thing I read in Bee Culture. Not so much because it's at the end – it's more like saving dessert for last. I can always count on Ed's combination of homespun wisdom and personal stories to leave me in a happy state."

Eugene Makovec, Editor, *American Bee Journal*.

"Ed Colby writes for the beekeeper. In each of his stories you might get a useful beekeeping tip, a mishap in the bee yard, a story of his latest adventure, or all of the above. He is quick and funny, as he blends his knowledge about bees with experience in life. His writing is a delightful mix of bee information and plain fun."

Dr. Meghan Milbrath, beekeeper and honeybee disease researcher.

"I delight in turning to the back of my monthly *Bee Culture* magazine and first reading *The Bottom Board*. It will be great to have this compilation of 60 of Ed's *Bottom Board* columns all in one place – and I can start at the front, middle or back.

Dr. Dewey Caron, Emeritus Professor of Entomology and Wildlife Ecology, University of Delaware.

"As I assemble Bee Culture every month I always like to get to the end. That is where Ed Colby's Bottom Board column is. He can turn a bad day into a good story every time with a smile."

Jerry Hayes, Editor, *Bee Culture*.

I have read Ed's articles and column for years in Bee Culture. His humor, storytelling, and descriptions of beekeeping pull at the common threads we all share in our appreciation for the honeybee. He personalizes his writing in a way that leaves you thinking it was written for you, the reader. This is a writer's skill to bring author and reader to the same moment around a familiar and shared experience. I thank Ed for making me smile and reminding me that most challenges in life are only as big as you make them.

Dan Conlon, President, Russian Honey Bee Breeders Association. Owner, Warm Colors Apiary, South Deerfield, Massachusetts

Foreword

I started as Editor of Bee Culture magazine the year after tracheal mites were discovered in the US. I was a new voice in the industry, and my first issue took off in some very different directions, much like the industry itself was doing. We added new, younger authors, covered exciting and often controversial topics, brought in better marketing information, more science and especially an editorial with opinions about all this that occasionally ruffled some feathers. That Editorial page was at the front, right after the contents page and was called the Inner Cover, because that's where a beekeeper starts working a hive. It also had the very last page strictly dedicated to humor or satire, making light of lots of things so the reader would leave that month's issue both smarter, and with a smile. That last page was the *Bottom Board*.

The wit, wisdom and humor of the Bottom Board attracted a steady following over the next decade or so, and several authors made contributions on occasional or even a regular basis. Then along came Ed Colby. And 20 years later he's still sitting on that Bottom Board, with a dedicated group of followers that'd be the envy of most state associations. He's in the top three of who gets read every month. That's saying a lot.

Let me give you some background on this guy. You'll find out a lot on these pages, but the details may escape you. He started with bees in 1996, and writing for Bee Culture in 2002. There are a lot of stories. Hundreds. Some of the best are here. But not all of them.

He's just a sideline beekeeper with maybe a couple hundred colonies in a good year. He made a living for 44 years working on the Snowmass and Aspen Mountain ski patrols, for the thrills. That's a story in itself that only gets a little attention. That job almost killed him once and he figured out Life is a gift.

Like a lot of beekeepers his little Darlins have taken him a lot of places, Mexico, Cuba, Ukraine, even Medina to visit me. And he's done most of the things a lot of the bigger guys have done. Bees to almonds, selling splits, beeyards in odd and grand places, and he's kind of a name dropper – Marla the scientist, Paul with the secret handshake, his gal Marilyn, Tina the tenacious, Meghan the skier. There's more. See if you can find them.

He's learned a lot of things the hard way. His dad told him to follow his heart, and Ed thinks because he did just that it's worked out OK for him. Most of the time anyway. His stories show a Beekeeper's Life worth knowing about. For both honey flows and dearths.

He finished one piece this way. *Our time on this good Earth is brief. Be at peace with that. But we beekeepers made the right choice. We understand what really matters. We took the road less traveled. Lucky us.*

We're lucky Ed took notes.

Kim Flottum

Author's note

I would be remiss to offer you this collection of beekeeping myths and tall tales (All true, to the best of my recollection!) without a word of acknowledgment and thanks to the people who helped and encouraged me along the way.

First, to Paul Limbach, commercial beekeeper, former boss, and gracious mentor, who showed me that not all wisdom is to be found in books, or on the internet. He taught me to keep bees, to fish, but most of all to listen.

To my former pianist-wife Linda Jenks, who put up with and even approved of my obsession with my little darlings. When she first heard the plaintive, descending cadence of caged queen bees piping in the kitchen in the wee hours, she turned to me in perfect awe. "Listen!" she whispered breathlessly. "They sound just like that bird in the Grand Canyon!"

To Tina Sebestyen, my vice-president, confidant and fiercest ally during my tenure as president of the Colorado State Beekeepers Association. She shared the secret to her remarkable beekeeping success: "I never, never, never give up!"

To my niece Pamela Singh, sister Patty Watkins and lifelong friend Frank Wetmore, and to Tamie Meck, all of whom cajoled and pestered me to stop procrastinating and complete this work.

To my dear friends, commercial photographers John Kelly and Irene Owsley, who provided photos for this book.

To former Bee Culture editor Kim Flottum, who took a chance on me.

And finally, to my philosopher-sweetheart Marilyn Gleason, who took a chance on me as well, who turned her life upside down to take up with a beekeeper.

We live under a nuclear sword of Damocles. In an age marked by political, economic, and environmental upheaval, and a seeming dearth of brotherly and sisterly love, the ancient and honorable craft of beekeeping still takes us down a path less trodden, to a world of wonder, and miracles. Maybe you turned off the main road a mile back. Good. Now here are some stories.

A bicycle ride

I've been riding my bicycle to town lately, for three reasons: To save the planet, reduce our dependence on foreign oil, and get a little exercise.

Coming out of Patti's Main St. Coffee House the other day, I mused that I might take a little two-wheeler side trip on my way home.

At first I thought I'd ride to the city limits, but when I got there, I felt pretty good. Maybe I'd turn around at a fork in the road a couple of miles farther on. But at the fork, for some reason I made a right, instead of a U-turn. Then I decided I'd stop at a nearby campground for a drink of water, but instead I just kept going.

We'd had a little rain, so when the road turned to dirt, it wasn't dusty. It was only a few miles to the end of the road.

And that was how -- like Forrest Gump when he ran all the way to the Pacific Ocean -- I finally wound up at Kelly's ranch.

Kelly told me a long time ago that he'd welcome bees on his spread, and I thought it might be worth a try. There's not a lot of alfalfa up here, but it might make a good staging apiary, where my little darlings could feast on springtime dandelions, before I move them to the high country. Maybe bees would flourish here all summer. You never know.

Kelly and I go way back. I was his straw boss when we cleared the original ski trails at Snowmass. When he came back from Vietnam, he made his fortune in ski photography.

Down at the house, I was greeted by his beloved dog, a cross between an Akbash sheep-guard dog, and an Australian shepherd. I like those Akbash.

Some places enchant you. There are a number of buildings on the property – the main house, a greenhouse, a sauna /guest house, another guest house -- all built by Kelly and his stonemason, who lives in one of the cabins. The buildings are not large, nor are they ostentatious. The accent is natural wood and stone. Everything looks like Kelly either made it himself, or he inherited it. There are photographs everywhere. You don't have to take your shoes off to go inside. And just the right amount of clutter makes me feel right at home.

Kelly and I sat outside and drank cool water from a tin cup. We talked about the good old days, and of course, bees. He was all for having them.

I asked about forage, and he pointed out a few fruit trees. I tried to explain that it really takes more than that. But mostly I was looking for a migratory location with dandelions. He assured me the valley turned bright yellow in May.

Kelly showed me a photograph of a bear that climbed an apple tree next to his house and got off on the roof. But I'm used to bears. All of my yards have them. I put up solar electric fences and cross my fingers.

I once built a fireplace for a gentleman with whom I had a separate business relationship. One of the laborers on the job had an unemployed stonemason friend who hung around some. He carried around a "portfolio" of photographs that featured his work.

I didn't think too much about it at the time. He might be the reincarnation of Michelangelo, but I had the job. I can swing a trowel. Plus the owner owed me. When he fell off the second floor and nearly severed his spinal cord, he went wacky. When I directed the carpenters to help me immobilize him, he got belligerent. He's used to being the boss, but I told him, "Don't worry. I'm in charge," He said some bad words. We held him down until the ambulance arrived.

Later, he shed a tear when he thanked me for saving him from paralysis or even death. No problem. I've been a ski patroller most of my life. This was just basic first aid.

After I finished the fireplace, there was some remaining stone work that for some reason wasn't supposed to get done until the following spring. But in May, when I asked when I could get started, I learned that that other stonemason had gotten the job.

The owner called it "a business decision." Naturally there were hard feelings on my part, more towards the owner than toward my competitor.

That was 20 years ago. I still see the owner from time to time. I sell him honey. We laugh and shoot the breeze like all this never happened.

I always wondered what happened to the other stonemason, and now I know. He lives at my old friend Kelly's ranch. He laid Kelly's beautiful stone floors, the stone walls, the "water feature" in the greenhouse. He's got the touch.

Kelly introduced us on my visit, and we shook hands like it was the first time we'd ever met. I wondered if he even remembered. I wouldn't have recognized him, but his name brought it all back.

Now I heard distant thunder. The sky looked ominous over the Flat Tops. I said, "Kelly, I'd better jump on my bike, before I stay too long."

Heading down the lane back to the county road, I reflected on meeting the mason with the portfolio again. It just didn't matter anymore.

What mattered was the rekindling of my friendship with Kelly, and a new home for my bees.

September 2010

A dark and stormy night

Last August as I drove out of my Flat Tops bee yard at nightfall, I heard my rear pickup tire hiss flat. Well, OK, I thought, I can deal with this. I can hold the flashlight in my mouth. That'll leave two hands free to change the tire. How hard can this be?

When I turned on my light, I discovered that I was dragging a three-strand section of barbed wire fence wrapped around my rear axle. I knew I had my work cut out for me. I put a short 2 by 6 on the ground and set the hydraulic jack on top of it. But when I tried to pump, nothing happened. It dawned on me that maybe my jack oil had leaked out.

Then the sky lit up, thunder boomed, and the rain came down in sheets. I was at 9,000 feet, 400 yards from a lonely county road, and 30 miles from home and hearth in Peach Valley. Linda would just have to come and get me.

I was in a meadow in plain view from the county road. My watch read 8:45. By 11 at the latest Linda ought to be worried enough to come looking for me. I thought about waiting in the truck with my dome light on, but when I saw three cars drive by on the county road, I decided to hike over there. More cars were bound to come by, and if they didn't, Linda would come before midnight, for sure.

It was as black as a tomb. I was halfway across the meadow when it hit me that the dripping wildflowers were soaking me to the waist, but by then it was too late to turn back. At an unlocked cabin by the road I stood on the porch out of the weather and waited in vain for a car to come by. When it stopped raining, I stood right in the middle of the road, hoping. More than once I thought I heard a rustle in the bushes. Far off in the distance an Akbash sheep guard dog made his rounds. He howled like the hound of the Baskervilles on the lonely moors, but I'm a grown man, and I don't get the creeps.

At midnight I gave up on Linda. I wondered if maybe she went looking for me at the other Flat Tops yard and slid off that nasty little stretch of four-wheel drive. There was nothing I could do about it in the middle of the night.

I went inside the cabin, but it wasn't a place you'd want to sleep. Mice had pretty well taken over. No electricity of course. There was paper and kindling in the wood cook stove, and I thought a fire might be nice. It wasn't that cold, but I was soaking wet. There were no matches to be found, however, and mine were back in the truck. The larder was well stocked with pork and beans, Spam, and flour.

I pulled a threadbare sleeping bag off the wall. For some reason the mice had never gotten into it.

The night in the truck wasn't so bad. I climbed into that sleeping bag and got a little rest. The moon came up.

At dawn I struck out for civilization. I had a dry bee suit in the cab. I didn't see any point getting it wet walking out to the road again, so I threw it over my shoulder and lit out buck naked through the mint and the coneflowers. OK, I had on a jacket, and shoes, but that was it.

I had my bee suit on a mile or two down the road when I heard Linda's car. She looked relieved. I said, "Where were you eight hours ago?" Two seconds later I wanted to eat those words.

She eyed me up one side and down the other, and all of a sudden she didn't look so happy to see me. "Are you crazy or what?" she said. "You know I can't drive at night. There were mudslides on the road. I called the sheriff, and a nice young deputy almost slid off the road looking for you. He knew about the Dodo yard. He finally made it up there and yelled for twenty minutes in the dark. Whatever possessed you to leave home at four o'clock with storm clouds rolling over the Flat Tops? You got yourself into this."

Which was true.

We didn't talk a lot on the way down. Down where the BLM land gives way to irrigated farmland, I said, "I saw the biggest coyote right here the other day."

Linda's head swiveled ninety degrees, and she looked me right in the eye, just as serious as any girl could be.

"I saw that coyote this morning," she said. "He was so big, at first I thought he was a deer. But he had a dog in his mouth. A black dog, or maybe a lamb."

Down the road a bear sauntered into the oak brush. Linda said, "Oh, my God. That bear is huge. You could have gotten eaten."

It didn't look like that big of a bear to me. I was about to explain to Linda that it was only "an average-sized bear" but then thought better of it. I was in deep enough already.

October 2005

You just have to be careful

I never would have gone back to Brazil just to see Africanized honeybees. I went to see Jimmy.

I didn't even hear a rumor of Jimmy for 42 years. In the sixth grade, he and I scrapped outside Rio de Janeiro's American School. When he wrote to me last winter, he wanted to talk about it.

On the plane down to Rio in August, I compensated for long-forgotten Portuguese vocabulary in a uniquely American way. I talked louder. By the time we landed, all the Brazilians in the forward section of economy class had heard about Jimmy.

They learned that after our little fistfight, Jimmy and I became friends. They found out that he bought the seventh grade class presidency for me with Baby Ruths, Juicy Fruit, and Parliament cigarettes – all smuggled off the ocean liner SS Brasil. They learned that I left Brazil after the seventh grade, and that now I had to fly to Rio for a weeklong school reunion to explain to Jimmy that everything is all right, forever and ever.

The Brazilians all said, "The crime in Rio is terrible. Beware of thieves." The nice missionary lady looked up from the Book of Romans and smiled. "If you're lucky, they'll leave you standing in your underpants," she said.

In Rio I heard stories, like the time bandits held up the big tunnel that connects two main parts of the city. "Everyone who lives here has been robbed," my Brazilian classmate Suelena said.

Through a friend of a friend of a friend of a friend I'd found Walter Gressler, a retired beekeeping instructor at the Escola Wenceslau Belo, an agricultural school in Rio. When I called him from Colorado, we arranged to rendezvous the following Wednesday. He asked what my main interest was. "Africanized honeybees," I said.

On Tuesday evening I dined at the Rio home of a former classmate. We got up from the table at 1 a.m. Jimmy drove me back to my hotel like a maniac. When I remarked that he'd just run a red light, he laughed. "Nobody stops for red lights at night. Too many muggings." By the time he dropped me off it was 3 a.m.

Four hours later, I hailed a taxi that immediately plunged into morning rush-hour gridlock. When we finally emerged from the snarl on Avenida Epitacio Pessoa onto a crowded freeway, I mentioned that I was running a little late. "I'll get you there as fast as I can," my affable driver replied. He tailgated, sped, swore, and cut off other drivers. My seat belt wouldn't latch.

At the Escola Wenceslau Belo, Walter and I walked through lush tropical forest to reach the bee yard. Although the grounds of the school encompass 84 acres – entirely surrounded by the city of Rio de Janeiro – this Africanized honeybee yard sat only 100 yards from a city street.

Next to the hives we ducked into a little classroom. Walter said, "Write this down. Never go into a bee yard alone. Also, Africanized bees don't like dark clothing, barnyard smells, menstruating women, swishing tails, flying hair, flying shovels, pounding hammers, perfumes, smelly people. Wear clean clothes, don't walk in front of the hive, don't move your hands over the hive, don't shake the hive, avoid quick movements, and use plenty of white smoke. Too much smoke is better than not enough. Never let your smoker go out."

He paused for emphasis. "Those silly little German smokers won't do the job. You need a Brazilian smoker."

His "Brazilian smoker" was about twice the size of anything you probably ever used. Walter stuffed it with wood chips and then green leaves on top to cool the smoke. When we opened the hives, Walter not only smoked the bees – he also laid down a thick layer of smoke all around the hive.

The bees themselves behaved like, well, honeybees.

I tried to help. Walter said, "Pick up the hive covers by their edges. Underneath there could be a spider or a snake."

"This bee is Africanized," Walter said. "But this is a Brazilian bee. She combines the European bee's natural disposition to make honey with the African tendency towards making lots of brood."

Africanized Brazilian colonies produce up to 450 lb. of honey per year in a country where migratory beekeepers find honey flows year-round.

According to Walter, Brazilian beekeepers face practically none of the serious problems endemic to American beekeeping. No foulbrood, no sac brood, no serious mite problems, no hive beetles. He did mention a poisonous pollen problem with a tree native to the Atlantic coast, but he said beekeepers got around that by feeding artificial pollen when the *barbatimao* tree blooms.

When I asked about Africanized bees' reputation for aggressive behavior, Walter corrected me. "Defensive, not aggressive," he said. "They only defend the hive. I don't know how it is in your country, but forty years ago, when these bees moved through Brazil, it was horrendous. They killed people and animals alike."

He explained that Africanized honeybees behave most defensively on the frontier of their expanding range. Once these bees crossbreed with local European bees, they become much more manageable. A "hot" Africanized queen might head a colony that habitually chased intruders a kilometer or more, but after repeated crosses with European bees, her genetically watered-down progeny might be inclined to pursue you, say, only a couple of hundred meters.

"We can live with these bees," he said.

Except he almost didn't. When Walter once tried to burn an overly "defensive" colony, he instead burned a hole in his veil, and survived 129 stings.

After we finished working the bees, Walter and I went back into the classroom. Walter put his hand on the back of my zipper veil. He said, "Did you know your zipper is wide open?"

Walter showed me around the school. We looked at hogs and laying Cornish game hens and hives of sting-less Brazilian bees that make a kilo of honey a year. Walter's easy way, his genteel hospitality and the strength of our beekeeper bond touched me. He gave me a day out of his life.

Late in the afternoon Walter and I walked to the busy freeway next to the school and hailed a taxi. "Take this man to his hotel in Arpoador," Walter told the driver, "and see that no harm comes to him."

The driver pointed indignantly to his picture on his taxi license on the visor. "Sir," he said, "I am a registered taxi driver in this city."

All the way back, I fantasized about being driven to some drug lord-ruled slum, where men in tattered trousers would take my money, and then . . . I shuddered, but my driver only wanted to talk politics. "George Bush, killer," he muttered in English all the way back to Arpoador.

Part of my heart lies in Rio. I grew up there. My feelings about this place echo Walter's when I asked him his opinion of the Africanized honeybee. "I'm in favor of her," he said. "You just have to be careful."

April 2003

17

In your ear

Back in April, a cold wind howled as we worked the honeybees at the Wallace Creek gravel pit. In an offhand way, Paul said, "There's a bee trying to get in my ear. She probably just wants to get warm."

Paul has shaggy ears, so I can understand how those ears might look like a refuge from the storm. Bees somehow get inside our veils all the time, but Paul never ripped his off to shoo that bee away. He never stopped working. He merely remarked.

He continued: "A bee could get stuck in your ear. You might have to get somebody to get her out with forceps." Paul loves to talk like this.

I had a honeybee up my nose once, but a bee inside my ear sounded downright distracting. "You're giving me the creeps," I said.

"Maybe the bee would crawl right inside your head," Paul said with a wry smile.

"She wouldn't get past my eardrum," I said.

Paul said, "A bee couldn't sting your eardrum, because she'd have to back in to do it."

I said, "Who mentioned anything about eardrums getting stung? I think any sting inside your ear would really light you up. You know, I'd wear ear plugs, but I can't hear that well even without them."

Paul said, "I think you have more important things to worry about."

Back in the truck, Derrick turned up the radio. He said, "This song really gets to me. Have you seen the video?"

I said, "No. I also can't make out the words."

Derrick said, "It's about this woman who kills her little girl."

"You're kidding," I said.

Derrick said, "First the girl hides her bruises at school. Then she ends up at the cemetery. You should see the country video."

"I'd rather not," I said.

Just then I thought I caught some of the words. I said, "'Name on a polished stone?' Did I hear that?" Derrick said, "Yeah, 'name on a polished stone.' That's her gravestone."

Derrick, who is 21, recently came back to work for Paul after a couple of years away. He rhapsodizes about how this job is better than his last one, managing a shoe store. I concur.

You couldn't ask for a better job.

Paul drives the truck. At first Derrick sat in the middle, out of deference to my advanced age, I suppose. But as soon as I made it clear that we're all equals, Bro, he kind of seized on the shotgun position. With the window down, he can smoke. You really can't smoke in the middle.

In the shop the next day I said, "Derrick, the name of that song is 'Stone Baby,' right?"

Derrick gave me a look. "'Concrete Angel,'" he said.

"Concrete Angel and Stone Baby kind of sound the same to me," I said. Derrick shook his head. Didn't I know anything? Paul chuckled at our morning banter as he headed out the door.

Just when I think I have all my friends convinced that I'm some kind of saint for keeping bees, somebody rains on my parade. The latest: Robin says she won't buy my honey anymore. I asked why.

"It's because you smoke out the bees, Ed," she said. "All you beekeepers do. I can't go along with that."

I said, "Robin, what are you talking about?"

She said, "You smoke out the bees to kill them so you can get their honey."

I said, "We give the little darlings a puff or two to pacify them. Who ever told you otherwise?"

She said, "It's in some information I got from an environmental group."

I said, "Robin, who do you believe? Me or somebody you don't even know? Did it ever occur to you that whoever wrote that maybe has no idea what they're talking about?"

She said, "Ed, I'll bring you the information. I'm pretty sure you kill those bees to get their honey. That's why I can't buy any."

At least my good customer Gail from the garden club thinks I'm some kind of saint. When she invited me to speak on honeybees at the club's August meeting, she indicated that the club customarily paid an honorarium. How much would I require?

When I threw her a number, she said, "My, that's about double what we customarily pay."

I said, "Well, you asked. I figure with research, rehearsal, driving time, plus a two-hour meeting, that's a bargain. I understand that the club is your volunteer work, but it's not mine. I appreciate the good you do, and I appreciate the opportunity to preach about honeybees, but Gail, I need a fair remuneration. I'm worth it."

I'll sell my talk the same way I sell honey. I'm going to charge what the market will bear.

She seemed to comprehend my point of view and said she'd float my proposal. Gail's the club president, so she has some pull. When she called back, she said the talk was a go, even at my "rather steep fee." I promised not to disappoint her.

In my talk I plan to include a discussion of the theory and practice of smoking out the bees.

Even though the other night I dreamed we won $3 billion in the lottery, the reality is that Linda and I will never be rich. Still, we both love what we do. We have our health (knock on wood). We have our good dog Spot. We have the bees. The universe looks OK from here.

Everything will be just fine, as long as a bee doesn't crawl inside my ear.

July 2003

Commercial beekeeper Paul Limbach shares a little knowledge with members of the Colorado State Beekeepers Association. Photograph courtesy of Marilyn Gleason.

Staying cool

"The king was in the counting house, counting out his money . . ."

My beekeeper boss Paul loves the summer dog days because he gets to start extracting honey.

When he estimates the net weight of his honey-filled supers in the honey house, multiplies it by the market price, and makes a calculated guess about future production, it gives him an idea how much money he might make this year. Paul doesn't talk about it, but we all know. With honey prices hovering in the stratosphere, in his mind he's already tarpon fishing in the Keys.

Summer dog days you ask? Isn't the honey flow over, and the honey about all in now? Well, it might be October as you read this, but as I write, it's late July and the tail end of the hottest Colorado summer anybody can remember.

The sheet metal honey house heats up above 100 during the day, and probably a lot hotter, although nobody actually checks a temperature. I guess you want a hot honey house, so the honey spins freely out of the comb. Paul cranks up the radio and works alone all day extracting. He sends the crew to work the bee yards from "Out south of Silt" to Rifle to Parachute to Meeker to the Flat Tops to Steamboat Springs.

Temperatures soar to the upper 90s most days. Of course none of the trucks has an air conditioner that works.

The July issue of *Bee Culture* ran a letter and photo from Dick Crawford of Morrissonville, NY. The picture shows three beekeepers in their beesuits and veils standing shoulder-deep in a swimming pool.

Dick wrote: "We jumped into the pool with our beesuits on. Then we went and took off honey. Our clothes kept us cool while the temperature was 95 degrees . . . I got the idea while working construction -- jump in about noon and work in cool clothes all afternoon."

Stay cool merely by staying wet? This is too simple.

On the way to the Zehner bee yard outside of Hayden, Derrick and I laugh about Dick's picture. Later, loading honey supers onto the truck, I feel so hot it occurs to me I might actually get sick. Afterward I park by the Yampa River.

Derrick says, "What are we doin'?"

I say, "We're goin' into the river with all our clothes on."

He says, "We are?"

You never saw a kid jump out of a truck so fast.

Arms outstretched to the side, I walk across the stony bottom in my long-sleeved shirt, jeans and Converse All Stars. Maybe some stern-faced preacher will push me under and baptize me. From waist-deep in the current I plunge head first into the riffle, and the cool river gently sweeps me away. When I come up for air I'm 12 years old again, and the world looks new.

Now we go to the river every day. Even Mark succumbs to its Siren song. He sits hunched-over on the bank, struggling to remove an upturned boot, just like the cowpoke in corny cartoons. Then he steps gingerly into the river, grinning, as the sun illuminates his ghost-white cowboy shoulders.

Swimming-with-your-clothes-on as an energy-efficient personal cooling system totally changes my outlook on hot weather. A noon plunge keeps me chilled for about two bee yards. When I'm almost dry I put on my spare pre-soaked shirt, and it's October again, no matter how hot the day.

Derrick and Mark don't buy into the wear-all-your-clothes, soaking wet idea. They take their shirts off when they go in. That's their problem. Mark's pretty careful in the river. Everybody knows cowboys can't swim.

Derrick basically works bees naked anyway. He generally wears shorts, shoes and a veil. That's it. He does stay pretty close to the smoker, and he sometimes says a bad word, but who doesn't? People stop alongside the road to take his picture.

I don't get it. I can't understand why everybody doesn't walk around fully clothed and dripping wet all summer.

The other day when I climbed out of the river I almost tripped over an old bison skull half-buried in the sand. I have the picture to prove it.

Right away I knew what it is. Broader and more massive than a cow skull, its short, up-curved horns gave it away.

Mark knows his western history and Indian lore. He said, "The Sioux used to come down here from Wyoming to chase buffalo." The skull had a hole in one eye socket, and we speculated about that.

Mark finds arrowheads and spear points all the time. Whenever we look together, he always finds stuff – at least chips -- but all I ever see are rocks and dirt and sagebrush. After I found the bison skull he kept saying, "That's an unbelievable find. How'd you get so lucky?"

Lucky finds never happen when you put your head down and grind it out. We drive right past this spot all summer. We could have almost seen that skull from the road.

I'd have never found it if I hadn't been swimming in the river. Thank goodness for hot weather, an air conditioner that doesn't work, and a picture in *Bee Culture*.

October 2003

Little Darlings

I used to be a wallflower, but since I started keeping bees I'm the life of any Valentine's Day party.

You already know what I mean, don't you? Once strangers learn I'm a beekeeper, there are no awkward silences. Folks start asking questions and look to me as the expert that I'm really not.

Right away I figured out that the title of beekeeper confers a sort of celebrity status, only without accompanying wealth. You see, bees intrigue people. Almost everybody thinks that bees are important to the environment, although they're maybe not sure why. They've heard that mites are a threat, although they're not sure how. They know killer bees are out there somewhere. And of course they fear bees. Who hasn't been stung?

Beekeepers know the answers to all their questions, but more significantly, they're different in that they are not afraid. How could they be? Somehow they go into a beehive – into that scary maelstrom – and pull sweet honey out of it. What could be more mysterious or wonderful?

Beekeepers connect to the land in the way all farmers do. They ply one of the world's most ancient and respected crafts.

I say milk this.

I knew I wanted to write, but I needed an angle. Why not bees?

Aunt Minnie came up with a name for my proposed column – "To Bee or Not to Bee." I live 55 miles from Aspen, and the best known, most widely read paper around here is the Aspen Times. The editor said, "What's with the name?"

I said, "Well, if I write about bees, it's 'To Bee,' but if I write about something else, it's 'Not to Bee.' The editor laughed. He said he'd give it a try.

Residents of one of the world's richest, most cosmopolitan communities now get to read about piping queens, evicted drones, crop dusters, jet-black aphid-spit honey, butterscotch-flavored rabbit brush honey, raiding black bears, bees at 9,500 feet, Africanized bees, and the always unpredictable twists and turns of any day working for my beekeeper boss, Paul.

I try to amuse folks, but I also want to educate them about bees and make them feel sympathetic to bees and beekeepers. I started calling honeybees "little darlings," and that became a sort of trademark of the column.

In June another Aspen Times writer, veteran columnist Su Lum, weighed in on honeybees and a certain local beekeeper. She wrote:

"Another peculiar thing about this summer has been an inundation of BEES . . . I started asking people if they'd noticed a proliferation of bees and they said, 'Yeah, man, a lot of bees this summer.'

"Two weeks ago I was reading Ed Colby's sweet column, 'To Bee or Not to Bee,' and damned if he didn't say he was trucking his bees up to ASPEN . . .

"I think we need another column from Ed Colby to clarify this matter. How many of the little darlings are you talking about? How do you transport bees to greener pastures? Do you lead them here like the Pied Piper, or do you bring a truckload of bees, and set them down somewhere? Where? Do you need a bee permit?

"After the initial shock at the plethora of bees, I really don't mind them. They do not divebomb or (so far) sting – they appear to be, indeed, darling bees, perhaps lethargic due to all the smoke.

"Speaking of the Pied Piper, what if you were raising darling RATS? If conditions were more favorable here, would it be OK to send them to summer camp in Aspen?"

Cute, huh? Initially I basked in the glow of this unexpected attention. "Hey," I thought, "People are reading my column!" But Su's urban point of view troubled me. She asked a leading question: how many bees? What was I going to say, "Oh, just a million or two, Su . . ."?

Su's message was clear. She'd never really noticed bees in Aspen before, but now that I had a bee yard in the neighborhood, suddenly the town was thick with them.

I don't want to paint with too broad a brush, but Aspen is a small town filled with big-city folks. Some will call their attorney before they'll phone a neighbor. I could picture it. A child gets stung by a honeybee, a wasp, a yellow jacket, whatever. She has an allergic reaction and gets rushed to the hospital. The parents need a scapegoat and a legal target. Is their case weak? Of course, but I might still need to hire a lawyer.

Then a friend from Aspen called. He insisted that the city suffered from a plethora of yellow jackets, not bees. I responded to Su in a column explaining the differences between honeybees and yellow jackets. I pointed out that the omnivorous yellow jacket makes trouble, not honey. I wrote: "Yet time and again people crucify the saintly honeybee for the sins of the yellow jacket."

I summed up: "Personally, I don't know a thing about any kind of plethora in Aspen. But go ahead and call it a plethora of bees, if you want, or call it a plethora of yellow jackets. Just don't call it a plethora of Ed's bees, because my bees don't live in Aspen. They live way out in the country, and they never go to town."

I immediately felt better. The Aspen dandelion honey flow was now finished, so I played it safe and brought the little darlings home.

February 2003

Aspen Mountain bee yard. Photograph courtesy of Ed Colby.

A cheapskate

On our budget, being a cheapskate helps.

Six weeks ago, Linda bought a new front door for the house. She's been talking about one for years, but last week she returned it. "I got $430 back," she crowed.

She's muttered about buying a horse and a Baldwin grand piano for ten years. She adores horses, and she's a musician by trade, so why not? We could afford both. We really could. But when it comes down to actually laying out the money, Linda won't pony up.

She's always been like that.

This summer I'm sharecropping 400 honeybee hives up by Steamboat Springs. This is a huge expansion of my little beekeeping operation, and I knew I'd need to invest in a real truck. I asked some guys who know trucks what they recommended. In January I started shopping for a 1995-99 one-ton 4WD Dodge Cummins diesel with a long-bed and manual transmission. I initially budgeted $15,000, but Linda immediately put up a squawk.

"Ed," she exhorted, "You don't know how this summer is going to work out for you. Fifteen thousand dollars! Trucks don't appreciate in value, you know."

Normally Linda and I drive beaters, but I wanted a reliable truck. I'd haul bees with it, and the last thing I needed was a breakdown on the road with bees on board.

I don't like to kick tires, so I use a car buyer in Boulder. He visits two auctions a week and comes up with some good deals. You tell him what you want, and he'll buy it for you wholesale, plus his fees. He knows car and trucks, and he'll buy a vehicle for you on your word alone, over the phone. I know and trust the guy, and this is the way I like to do business.

After I told him I could go $13,500, plus fees and taxes, Linda went ballistic. I reluctantly called him back and told him to drop it to $10,400, which would bring my bottom line to around $12,000. I told him I really only needed a cab and chassis, because my former beekeeper boss Paul said he'd build me a custom flatbed for "under $1,000." So we're really talking $13,000, but I decided to leave the custom bed off-budget, in order to buy as much truck as I could get away with.

For a while I thought I had Linda on my side, but I should have known. The problem was, my buyer couldn't find any Dodges in my price range. There just aren't a lot of used Dodge diesel one-tons out there. I expanded the search to Fords, but still no luck. February slipped away, and now time was running out. Paul said that after March he wouldn't have time to build me a flatbed. Every time I mentioned the truck search to Linda, she made it clear she

---

didn't really approve of such a large expenditure. Still, she didn't come right out and say "no."

I still wasn't smart enough to shut up. When I mentioned to Linda that Paul was unloading his old bee truck fleet, she said, "Well, why don't you buy one of those?"

"They're junk," I said, "He's selling them because he's tired of spending money to repair them. That's exactly what I don't want."

Linda said, "Well, just because you spend $12,000 for a truck doesn't mean you won't have to pay to fix it, too."

You have to admit she had a point here.

Paul's 1983 one-ton 4WD Ford has a new automatic transmission, new tires, and 40,000 miles on a rebuilt 460 engine. It's an E350 chopped off van with a flatbed. Weirdest truck you ever saw. The sign in the window read "$2,500." The cab on the driver's side is creased like the truck's been tipped over. The paint is mostly gone. It wasn't exactly showroom clean, but Paul had thoughtfully removed all of the empty pop cans, hunting regulation books, fishing gear, bee veils, bee suits, and leftover lunches from the cab. When we got out of the car for a closer inspection, I groaned, but Linda said, "That's a good looking truck. And hey, the price is right."

I drove this truck when I worked for Paul last summer. It is what it is. It wasn't what I was looking for, but I knew the vehicle, and I'd be buying it from Paul, who at 56 still hasn't figured out how to tell a lie. Plus, buying a cheap rig would have two immediate paybacks: the warm glow that comes with landing what arguably is a really good deal, along with short-term marital bliss. You can't put a price tag on the latter.

I'm a free man, but I've been married long enough to understand my options. Every action has its consequence. Did I want a carrot or a stick? Paul agreed to come down $200, and we sealed the deal over the phone. I felt a sense of relief that this saga was finally over.

As Linda, Paul and I walked down Paul's driveway to get the title, I said, "How's that latest Dodge truck you bought working out?"

Paul said, "The transmission went out two weeks after I bought it, and it cost $3,700 to fix."

Linda looked at me knowingly. Maybe she gave me a wink. "See?" she said, "You were smart not to buy a Dodge."

May 2004

Down in a ditch

When we moved to the country nine years ago, I convinced Linda we needed a good watchdog. I never thought about a good bee dog.

I asked at the pound what kind of dog Spot might be. The kid from the honor camp prison who ran the place said, "I know his pedigree. His mother was rottweiler and lab, and his father was cocker spaniel."

"Impossible," I said.

"She must have been down in a ditch," he smirked.

I told him I was looking for a medium-sized dog with a mellow big dog personality. He said, "This is your dog. This little guy is great."

When he opened the cage door, a skinny black dog bounded out. He ran circles around us, then leaped into the arms of his only friend in the world – the kid. Spot weighed maybe 40 pounds. You could see the cocker and lab. Forget what I said about what I was looking for in a dog. I was instantly smitten.

"He's less than a year old," the kid said. "I adopted out his mother yesterday."

Linda came along to the pound to be sure I didn't bring home a silly little poodle, or maybe a wolf. A lifelong cat fancier, Linda loathed dogs. She looked disgusted as I admired Spot and scratched his ears. On the way home, she drove, and Spot rode on my lap. When we got to the house, she said, "The dog sleeps outside." That first night Spot and I slept in the detached studio in the backyard. He never left my side. The next night we slipped into the bedroom.

It wasn't all roses. We discovered that Spot had a temper and liked to bite. Not big rottweiler bites, but the cocker spaniel nipping kind. He could also give you that unnerving blank rottweiler stare. As he began to morph into a dominant male with an attitude, I began to have second thoughts.

One morning while I was shaving, I told Linda, "Well, if Spot doesn't work out, we'll just have to put him down."

Linda turned on her heels. "We don't kill animals at Colby Farms," she said fiercely. At that instant she bonded with Spot, and forever after became his advocate and protector.

I talked to some dog people. Heeding their advice, I got Spot neutered, and we enrolled together in obedience school. He came through both the operation and the training with flying colors, and his behavior improved markedly. He was always a good boy in class, but the teacher never liked him. I never liked her. Maybe it showed. She made the mistake of saying, "Never adopt a pound dog. You don't know what you're getting."

More than anything else, Spot likes to ride in the truck. He always used to go with me to the bee yard. The bees themselves didn't seem to interest him, or bother him. That was then. Now he's a bee snapper.

I owned a Rambo hive . . . you know the kind. Those Carniolans belied the placid reputation of their race. Nothing I ever did was right for these gals, and they let me know it. They didn't like to be messed with, but they were excellent honey producers. I learned to put up with them.

Disaster struck on a warm, cloudy, April day, just before the dandelion honey flow. Spot lay by the truck contentedly watching as I checked honey stores.

A couple of hives flirted with starvation, but the Rambo bees waxed fat with hoarded honey. The little darlings appeared industrious, prosperous, placid, even. They had way more honey than they needed. Their upper brood super lay packed tight.

 It seemed a reasonable act – to rob the rich to feed the poor. But, like human socio-economic tinkering, this move proved fraught with controversy. I lifted a honey-laden frame from the Rambos, gave it a shake, and instantly became enveloped in a maelstrom of angry bees.

I felt like an American tourist at an al-Qaida pep rally.

Finding myself the center of the bees' attention and the object of their contempt, I stepped back from the hive. Regrettably, this retreat moved me closer to the truck, and to Spot. Half the bees that were attacking me instead vented their fury on the dear boy.

Spot snapped and yelped and barked and cried, and he ran and he ran and he ran. The bees followed him 150 yards out into an alfalfa field, where he snapped and howled some more. I hesitated for an instant, because I had my own angry squadron of kamikaze dive-bombers, and I knew if I went to Spot's aid, it would only compound his predicament. But I had no options, so I sprinted over to him and began smashing bees as fast as they landed on his thick, black coat.

Finally the attack subsided, and we dashed back to the truck. We somehow got in without letting in any bees, put our tails between our legs and headed home.

Spot's life changed forever. Now he can't tolerate a bee anywhere around him. Spring is the worst. Bees cruise the yard looking for a flower, and sometimes they'll buzz Spot while he naps on the grass. First he snaps. Then he runs to the kitchen door. But once the honey flow starts, the bees pretty much leave him alone.

The only other bad time is when I move bees back home after the crop duster finishes here in Peach Valley. They spend that first day home exploring, orienting, careening about, and they can make some mischief. I learned to leave Spot inside.

OK, maybe Spot isn't so great around bees, but how many dogs are? He still has an important job guarding Colby Farms, one he takes very seriously. He just gives the home yard, and especially the electric bear fence around it, a very wide berth. But that's another story . . .

January 2003

Spot and pollen

I'm always looking for the miracle medical cure. I prefer that it be dramatic, as in a 48-hour recovery.

Last winter Doctor Al put me on a big dose of niacin and folic acid to lower my cholesterol. The cholesterol-reducing results were ultimately inconclusive. But two days after I started taking this stuff, my worn-out old ski patrol knees stopped hurting for the first time in 20 years. They haven't bothered me since. I told this story to two or three doctors, but they just shook their heads and became strangely quiet. None of them actually told me that I was nuts, but I knew what they were thinking -- "First the body goes, then the mind . . ."

Spot suffered from allergies all summer. You never saw a dog so pathetic. A few years ago I treated him with honeybee pollen, because it had helped my own seasonal allergies, or it seemed to. The problem with miracle cures is that you never know for sure. The last time I tried pollen on myself, I was sneezing 50 to 100 times a day. The pollen gave me some relief. But this was pretty much at the end of my allergy season, so who knows? Maybe I'd have gotten better all on my own.

But I even had the veterinarian intrigued when I suggested giving pollen to Spot. Our vet is a kind and dedicated young man who is also a little goofy the way people are who maybe smoked too much pot in college. "Wow, pollen might work," he said brightly. "Try it and let me know."

But alas, no amount of pollen gave Spot any relief.

This past summer was one of Spot's worst. In September, at the height of his misery, the dear boy chewed his feet. He rubbed his snout on the carpet. He scratched his face. He covered up his eyes with his paws. He kicked his ears until they bled.

A self-appointed guardian of all animals, Linda was beside herself with sympathy. "He's tormented," she said over and over. She didn't approve of the medications the vet recommended. But when I suggested that we again try treating Spot with pollen, she threw up a firewall of resistance.

"He could have an allergic reaction," she said. "Like Russ and that poor person who bought your pollen at the fruit stand in Aspen. They both wound up in the hospital."

The Russ story was old news, but I learned only last summer about the fruit stand incident. I'd sold to the stand owner for years. His place of business was just a half-mile from my bee yard. But when I inquired last June if he wanted more pollen, he balked.

"One of my best customers almost died from that stuff," he said. "I warned her to try just a taste on the tip of her tongue, but people are gonna do what they're gonna do. Man, I can't take on that kind of liability."

I wasn't worried about Spot having an allergic reaction. I was more concerned that his condition was driving him crazy. When I gave him pollen previously, it was in the spring. Now it was fall, so this would be different pollen. I figured it was worth another try. At least one thing was certain: Spot wasn't going to sue.

I tried to reassure Linda. "Pollen's not poison," I said. "It's just that some people are allergic to it. Some people can't eat peanuts. Or shellfish. But most people can. Spot's had pollen before. He'll be fine."

"Animals can't talk," Linda said. "They can't tell us when they're in pain."

"Sure they can," I said. "I'll keep an eye on him."

Linda gave me a look. "No!" she said, and I could tell she meant business.

I knew better than to argue. But I like to live dangerously. I began giving Spot rabbit brush pollen on the sly. Rabbit brush is ubiquitous around here. It was in full bloom, and I suspected it might be the source of Spot's agony.

I didn't want to get caught, but it was a risk I freely accepted. I did it for Spot. He's a good boy.

Spot does not particularly care for pollen. He favors hamburger, or, even better, buffalo burger. So I'd take a bite-sized burger ball, make a little depression in it, and fill it with pollen. Spot did look at me funny when he inhaled his treat. As in, "Jeez Louise, what was in that, anyway?"

He did not have an allergic reaction.

A few days later, Linda said, "Spot's a lot better. Have you noticed?"

I said, "It must be the pollen."

For once Linda was speechless. She paused for the longest time. Then she said, "Well, he's definitely better."

The dear boy was lying peacefully at my feet. He wasn't twitching. I knelt down beside him and stroked him gently. I felt like I'd done a good thing. I also felt vindicated, but I didn't think it would be too smart to rub it in. Not yet. I said to Linda, "I think it's a miracle."

"I think he's cured," Linda said, her eyes shining.

When I stood up, my knees creaked, but they didn't really hurt. Not like they used to.

January 2006

A problem nobody else has

When Pitkin County Commissioner Mike Owsley asked if I'd pick up a half-dozen nucs at my neighbor Paul's and drop them off at Woody Creek, I said "Sure." I was going that way anyway.

The Owsleys and I go way back. Thirty years ago when we were neighbors in Leadville, I'd leave my dog with Ann while Mike and I drove to work together. Later, I lived in a sheepherder's wagon on their Woody Creek property, up by Aspen. They never charged me rent. They had me over for supper a lot.

Today I live 50 miles from Woody Creek, but this spring I had bees next to the Aspen airport. Mike and Ann's seven-acre spread was practically right on the way.

Until a few months ago, "gonzo" journalist Hunter S. Thompson lived at the other end of Mike's driveway. It might sound like I'm namedropping. I just want to give you an idea of the neighborhood. But Woody Creek is fairly understated, for Aspen. This is rural. They aren't all monster homes.

Mike's had bees practically as long as I can remember. But they generally don't make it through the winter, and every spring he buys nucs. Mike doesn't take any of this too seriously. At his daughter's wedding reception at their place a couple of years ago, he caught me heading over to the bees. "Don't go there,' he said. "It's too depressing. I never got the bees out of the nuc boxes, and they all swarmed."

The morning I dropped off his bees, Mike said, "I've got some frames you can have," but they turned out to be unassembled wooden frames, and I politely declined his offer. I came to beekeeping only a decade ago, and I'm a confirmed molded-plastic frame guy. I've never assembled a wooden frame. I've repaired a few. I suppose nailing frames would make a pleasant pastime for someone with time on his hands, but that's not me.

Mike and I chewed the fat some, like old friends do, and it was late in the morning when I finally arrived at my Aspen yard. Looking north from the hives, dandelions carpeted the meadow golden as far as the eye could see. I went right to work making nucs.

After a time, I heard a roar, sort of. Bees streamed out of the hive next to the one I was working, but it wasn't quite that jet-taking-off roar you associate with swarming. I think maybe this was because bees had to exit through a pollen trap, and they couldn't get out fast enough to all take off at once.

I couldn't tell where they went. They were flying everywhere and seemingly not headed anyplace in particular, so I just went back to work. There's just so much time I can spend chasing swarms. It was maybe a half-hour later that I noticed a cluster of bees hanging on an alder. I tried to bend the tree so I could shake the little darlings into a super, but the trunk was too stout. Undeterred, I put a comb of honey in a bucket and used a stick to hang it onto a limb next to the swarm. When the bees started heading into the bucket I thought, "Perfect," and went back to making nucs.

Pretty soon I noticed a steady stream of bees heading back into the hive that had swarmed. I thought, "That's weird," but I didn't really think too much about it. There's a lot I don't understand about bees.

After I made my nucs, I plugged the nuc boxes and loaded them into my pickup. I was in something of a rush because I had to pick up Spot at the vet's 40 miles away in Glenwood, and they close at 4. I had to deliver some honey in Carbondale on the way.

When I went over to collect my swarm – to my astonishment -- the bees were gone. Maybe the queen never made it out of the pollen trap, and the bees went back to be with her. Maybe the queen had stayed behind. I don't know.

Headed down the highway I noticed bees leaking out the back of my truck. I stopped and threw some netting over them, but I had so much junk in the truck bed that I didn't do a very good job. I didn't have time to do a very good job. Not if I was going to deliver honey and get to the vet's by 4.

Predictably, in Carbondale I got stuck in traffic. I was still losing bees. People stared, but nobody screamed or got stung.

I missed a turn looking for my honey customer's house. When I finally found the street, I realized I'd neglected to write down the number, so I just started knocking on doors. I found him on the fifth knock.

At the vet's office in Glenwood I parked in the alley. A woman with a Chihuahua and one with a Pomeranian were ahead of me. The receptionist was explaining a bill to the woman with the Chihuahua when I interrupted. "Excuse me," I said, "but I have a problem nobody else in this room has." That was a conversation stopper.

The receptionist eyed me closely. "What kind of a problem?" she said.

"I'm leaking bees out of my truck," I said. Could I pick up Spot and run?"

"Leaking bees . . . Oh my," said the woman with the Pomeranian. "You do have a problem nobody else has."

The receptionist laughed and said, "Didn't she tell you? Your wife picked up Spot two hours ago."

September 2005

A trip to Garfield Creek

A convergence of events led to my being "contacted" by the local police last May.

It started with the crop duster. He called to say he'd be in Peach Valley "any day now" to spray for alfalfa weevil. He generally sprays Furadan – an organophosphate insecticide highly toxic to bees – about the time the dandelions here in the Colorado River Valley are finished. I customarily move my honeybees out of his range. This time I took them to safety up Garfield Creek.

There was actually a silver lining to the bee move. On Garfield Creek the dandelions were in mid-bloom, so their new location gave my little darlings a second feeding.

I'd moved all the hives but one – a colony that bounced back from American foulbrood last year. I decided to keep it separate from my other hives, so I left it at home. I'd take my chances with the crop duster.

Then my neighbor Butch called to say we had a bear on our ditch. Well, great. By day the bees would face toxic chemicals. By night, a bear. But I could always put up some electric fence.

Later that same day I noticed that the bees had discovered some honey supers in the garage, which has no glass in the window. I'd just had 30 packages parked right outside that window and no problem. Now, with only one hive left on the property, bees were making mischief. I guessed I'd just have to repair the window.

Then suddenly I decided enough was enough. I said, "Linda, I'm taking this hive to Garfield Creek tonight and not worrying about it anymore. You're coming with me."

Linda is always stressing out about something, and she uses this as an excuse to thwart my plans.

"I can't do it tonight," she said.

"Why?" I said.

"Because the church "Cabaret Night" is in a week," she said. "Nobody is practicing. I'll never direct another production like this again."

"You always say that," I said. "Then you always pull it off, and when the audience goes wild, it's all worth it."

"This is my last Cabaret Night. I don't know why I got myself into this," she said.

"Well then take a break and go with me," I said.

"Stop bugging me," she said.

A while later she lost her billfold. When I found it for her, I said, "Now you owe me."

She said, "OK, what?"

I said, "You have to go with me to take the bees."

God fashioned no place on Earth lovelier than Garfield Creek at twilight. The jumble of low mountain peaks looming over the Colorado Valley looks like a scene from one of those dramatic Japanese nature paintings. Linda held the flashlight as I carried the hive across the ditch and set it next to its sister colonies inside the solar electric bear fence.

On the way back home I took it easy, as we savored the last lingering shards of celestial crimson. Linda instructed me to brake for Bambis and bunnies. You feel like you're in a Walt Disney movie on that road. Suddenly Linda threw her arm against my chest. "Stop!" she cried. There's a little rabbit over there!"

But basically Linda always calms down in the truck. She stops muttering about George Bush and Donald Rumsfeld. She folds her arms and sits there in a sort of altered state. You couldn't imagine a more contented person.

Back in New Castle, I stopped at the convenience store for a quart of beer. At this place you have to prove your age, even if you're an old beekeeper. I remarked to the girl behind the counter that maybe this policy went a little overboard, but I couldn't get a rise out of her. She solemnly said, "We check everybody" -- as if this were some provision of the Patriot Act.

Former New Castle mayor Pete Mattivi is pushing 100. I said, "You'd check Pete Mattivi?"

"We check everybody," she said without cracking a smile.

When I turned left onto Main Street, there wasn't a headlight in sight. Linda said, "That's a police car parked over there."

When the car's lights flashed behind me, I thought, "Now what?"

The officer informed me that he was "contacting" me because I had turned without signaling. He repeatedly called me "Sir" in a way that was genuinely polite and not at all condescending. I did not point out to him the absurdity of signaling at an empty intersection. I know when to shut up.

When he took my license and registration back to his car, I said to Linda, "Sunday night in New Castle. He must be bored."

Linda said, "Ed, he's very polite. He's just doing his job."

I suppose deep down I look at cops as authority figures with guns and an occasional mean streak. I'm sure this comes from getting harassed for hitchhiking and certain other youthful indiscretions. But Linda sees the police as the guardians of civilization. This is evident in the way we talk. I call them "cops." Linda always refers to them respectfully as "police officers."

After cautioning me to signal next time, the officer gave me his personal card, "so you'll know who contacted you." I had to admit that was a nice touch.

As I started to turn back onto that empty street, Linda said, "You'd better put on your turn signal."

September 2004

A complaining neighbor

Every beekeeper's worst nightmare is the complaining neighbor. I should know. Before I became a beekeeper, I was the neighbor.

In 1991, when our adjoining neighbor Joanne said she planned to keep honeybees in her townhouse attic, I claimed I was allergic and might die if a bee stung me. To my amazement, Joanne, who is a nurse, accepted this without question and said she'd find another place.

I lied to Joanne because our townhouse was for sale. Her attic window -- the one she proposed to use as a bee entrance -- was only a few feet from our own. I worried that bees from her attic would come over to our place and scare off prospective buyers.

We had problems enough unloading our home. It languished on the market. Bargain hunters pitched us low balls and expected us to finance the deal.

The gentrification of Carbondale had already begun, and one nearby neighbor thought the community in danger of losing its blue-collar character. He wrote letters to the paper advocating a junk-car fleet at the edge of town so that fancy-pants folks would shudder and stay away. He also practiced what he preached. He apparently never threw anything away, and his own yard morphed into a sort of Smithsonian of junk.

This guy – a city councilman, believe it or not – had done me a favor or two over the years. Plus, I generally don't mind a certain amount of stuff lying around, so under different circumstances I might have found his yard merely amusing. But as our unit sat unsold for two years, my good neighbor's eccentricity wore on me some. Then Joanne announced her honeybee project. Forgive me for losing it. Maybe you would have, too.

You have to admit that Joanne had an ingenious idea – keep your bees at home, but out of the backyard where neighbors would be sure to complain.

In hindsight, I suspect Joanne's attic would have actually been a poor place to keep bees. Those south-facing attics got hot, even in winter, and warm bees fly – a fatal mistake on Christmas Day in Colorado.

Anyhow, for her 40th birthday, her boyfriend John gave Joanne the complete beginner's bee outfit – a couple of hives, a couple of books, boxes of unassembled frames, foundation, a hive tool -- everything. At her birthday party she celebrated by donning her new bee suit, gloves, and veil, and dancing around the yard merrily "smoking" her guests.

The tequila flowed. After awhile nobody even noticed that the birthday girl was wearing a bee veil and coveralls. By midnight Linda and I had had enough and went home. We put in earplugs before we went to sleep. When we stopped by Joanne's the next day, the place looked like a frat house the morning after. John was still asleep in the bathtub. And all Joanne could talk about was her bees.

She put her two hives next to an alfalfa field outside of town. Sadly, the little darlings never made it through the winter, and Joanne's long-term dedication to beekeeping never matched her initial enthusiasm. She moved on to other pursuits, but her brief foray into that ancient and honorable craft changed my life forever.

Of course the townhouse eventually sold, and Linda and I found paradise in Peach Valley. Sometimes we call it "No-Peach Valley," because in the spring Jack Frost normally takes the whole crop. The first year we harvested sweet cherries – big Bings and Lamberts – but then for a couple of years we got none. At first I blamed the frost. But you see, the same weather that brings spring frosts – snow followed by clearing skies and cold temperatures – also keeps bees in the hive.

I began to suspect that maybe it wasn't so much those 29-degree nights as poor pollination that lay at the heart of the no-cherry problem. I decided that instead of relying on feral bees, or somebody else's bees, I should put my own beehives right in the orchard. That way if there were any break in the weather during the bloom, my trees ought to get pollinated.

Normally when I get an inspiration like this, I procrastinate until it goes away. But this time fate intervened. On a snowy January night, I bumped into Joanne at a hot springs pool. I seized the moment, sort of.

First I came clean. When I confessed my fib about being allergic to bees, Joanne only laughed and scolded me. After all, it had been a few years. Then I pushed my luck and inquired if she still had her beekeeping equipment. She said yeah, maybe we could work a deal. We left it at that.

A year later in April I called her. She was in a big hurry to go somewhere. She was a bit agitated, you might say. She said, "I've got a new roommate moving in tomorrow, and I need some attic space. I was going to call you. You can pick up that bee stuff, or I can haul it to the dumpster. But I need it out of here by tomorrow."

That's how I got started beekeeping. All I had to buy were the bees. Now I never have to worry about pollination. And you know what? Linda and I still hardly ever get sweet cherries, but we get honey every year.

December 2003

41

Winter comes to Peach Valley. Marilyn Gleason photograph.

Marilyn Gleason photograph.

A good strong west wind

Linda has this recurring dream about us dumpster diving to make ends meet in our old age, which isn't so far down the road. Linda and I both somehow managed to avoid career paths that might provide even a humble pension for our golden years. Now we feel like two little squirrels who played all summer and never stored away any acorns.

Linda and I aren't insider traders, so let's not talk about the stock market or our IRAs. The other day we decided to look for a rental property that might provide an income someday when I can't lift a super, and we're reduced to relying on Social Security and a battered shopping cart.

We tried to pick an honest, no-nonsense real estate agent. When we thought we'd found one, he and Linda went looking at fixer-uppers in Rifle.

Linda asked the realtor, "Why is every house on this block for sale?"

The realtor said, "I don't know. They just are." Later, when Linda called the city, they said the block was in a flood zone. Realtors! Boy! You have to watch those guys.

The realtor said to Linda, "Oh, and what does your husband do?"

Linda said, "He's a professional ski patroller and a beekeeper."

The realtor said, "You know, I used to work with bees. I always found moving bees fascinating. You just put the hives on the truck, and the bees swarm and follow along right behind!"

OK. Conversation is an art. Maybe I exaggerate a little sometimes, too. Maybe we all do. But who does this guy think he's talking to?

Probably no aspect of our venerable craft so intrigues the public as the simple act of moving bees. It intrigues me, too. You load a few million of the little darlings onto the back of a truck, and you head on down the highway. Hopefully you keep it on the road. Can you even imagine a bee truck rollover?

Colorado doesn't have a law requiring netting your bees in transit, and I don't own a net. I admit I hang it out there when I just load and go. I rationalize this because I don't travel very far, and I always leave at night, or early in the morning, or when the weather's rotten. Still, bad things can happen to anybody.

A few Decembers ago, Little Nic helped me move a pickup load of bees from Crystal Springs back home to Peach Valley. It was some job in the snow. When we bypassed Glenwood Springs, I stopped at a stop sign, but right behind another vehicle. When the car ahead of me started up, I followed, but without stopping again directly in front of the sign. Right away I saw flashing lights in my rearview mirror.

When the cop got out of his car, I noticed that he looked like a nice young man. I said, "Nic, watch this." As the policeman approached the truck, I leaned out and said, "Good evening, officer. You're not allergic to bees, are you?"

The cop literally jumped back when I said this. Regaining his composure, he came up to my window.

"What do you mean, 'bees'"? he said.

I said, "I'm hauling some in back, and they might be a little stirred up, even though it is the dead of winter. You just never know," I said.

"Whose bees are they, and where are you going with them?" he said.

I said, "They're mine. I'm taking them home to my place in Peach Valley. Is there a problem, Sir?"

The cop sized up the situation, and there wasn't a lot he could say about the bees. Plus I can turn on the charm when I need to. I apologized for my poor driving. I'm a courteous detainee whenever I get pulled over. The young man issued me a stop sign warning, and I thanked him for it. Little Nic chuckled all the way to Peach Valley.

Veteran Aspen Times columnist Su Lum once openly and publicly questioned my hauling bees to Aspen to feast on lush May dandelions growing on America's priciest real estate. What the heck did I think I was doing, anyway? She was actually pretty cute about it. She was mostly just amazed at the audacity of my undertaking. But what seemed to most intrigue her was how I got my bees up to Aspen in the first place. Did I lead them to greener pastures "like the Pied Piper?"

Maybe Su got this from our realtor.

I once told a shirttail relative about shipping bees to California to pollinate the almonds. He's a very sincere guy who would never pull your leg, like I generally do. He said, "OK, so let's say you've got all these bees out in California. Fine. Now how do you get them back to Colorado?"

I paused for just an instant, then said, "You wait for a good strong west wind. Then you turn them all loose. They're home in about three days."

My friend took this in without even cracking a smile. I looked him right in the eye. Why shouldn't he trust me? Just about everything else about bees is astonishing. Why wouldn't he

---

believe that, given a chance, good bees would head straight for home, like Lassie?

But then I broke the spell. I always do. You see, even though I'm a big kidder, I can almost never keep a straight face.

June 2004

Erving

After my arthroscopic surgery this summer, the surgeon said. "Your knee looked a lot worse than I expected, but I did the best I could!"

This wasn't the report I was looking for, but I had to appreciate his candor. Why buoy the patient with a false sense of optimism?

I was flat out of commission for awhile. Paul pulled my Flat Tops honey for me when he picked up his own Flat Tops honey in late August. I still needed to get 40 colonies off the Flat Tops and down to Silt Mesa for the winter. I wanted to do this sooner, rather than later, because you'd better not count on good weather at 9,000 feet around here after Labor Day. But with my bum knee, I couldn't hand-truck them up a ramp onto my flatbed. I'd have to hire somebody.

My tenant helped me one evening after work, but I underestimated the time this move would take, not least because I forgot how long it takes to staple beehives together. We worked until dark but only managed to load 19 colonies. You'd think I'd have remembered to bring a flashlight.

So I still had 21 hives at two locations to pull, hopefully in one evening. This time I decided I needed somebody with some real mojo to help, because it's a wicked push up that ramp at the Lettuce Patch yard. The hives were plugged double-deeps with a pollen trap and one mostly empty honey super. When I put out the word to Paul's crew that I had some generous overtime available, 20-year-old Erving jumped at the opportunity.

Erving is not a big kid, and I barely knew him, but Eric whispered that "Erving carries five medium supers at a time." Five? He sounded like my guy.

I told Erving I'd pick him up at Paul's after work. We'd load the little darlings that evening and unload first thing in the morning. Erving lives down the road a stretch, so I told him he could sleep at my place. I'd feed him supper and breakfast and pack his lunch. And I'd get him at work at Paul's by 8 a.m.

"So the job comes with meals and a hotel room!" he said cheerfully.

"Yeah," I said. Hey, whatever it took. I needed those bees moved.

I was in a tizzy when Nick and Erving rolled in 20 minutes late for our rendezvous. We were on a tight schedule, and I didn't want to finish this job in the dark!

Erving got out of the truck and said, "Just a minute. I gotta help Nick back up to the dock."

"Forget it. I've gotta go!" I snapped. I climbed into my truck and was about to pull away, when Erving jumped into the cab. He never said a word. He didn't have a toothbrush, a pair of socks, or even a jacket. He looked like a waif sitting there with absolutely nothing but a bee veil for an overnight kit. I wanted to laugh, but I didn't.

We had nearly an hour's drive to the bee yards, and my radio doesn't work, so we got to talking, or at least I did. I found Erving's story of his youth in the big city captivating, although I had to practically pry it out of him. A native-born American, his parents came from Nicaragua to escape the Contra insurrection in the 1980s. He grew up in New York, Miami, and "Jersey." This streetwise kid talks like he's from the Bronx. If he auditioned for a part in "West Side Story," he'd probably get it.

He wound up in Colorado when his dad, who owns a small fleet of trucks, found work as a subcontractor for the natural gas rigs.

He said his family owns some land in Nicaragua. There'd been talk about raising cows on it. When I asked if he thought he might like to go down there, he said, "Yeah, but my Spanish is pretty rusty."

I'll bet it's still better than mine.

We knocked out the Lettuce Patch yard pretty fast. Erving pushed the hand truck and I pulled with a rope. He wore slippery shoes, which made for some interesting moments.

We tied down the last of the hives at the Dodo yard with 5 minutes of daylight to spare. Just down the road, Erving spied the biggest bull elk. He said, "What is that? A moose?" I told you he's from Jersey.

Back at the house, Erving smelled the crock pot buffalo stew as soon as we walked in the door. "I hope you're not a vegetarian," I said.

After supper, I said, "I don't have a TV, but you'd better get some sleep, anyway."

I didn't hear him stir all night. I never woke him until breakfast was in the pan. He was still not fully awake at 5:45 when I dropped three eggs onto his plate. I said, "That doesn't look like enough" and gave him three more. He just smiled.

I packed him a quart of stew and some crackers and grapes and apples for lunch. Otherwise he'd be eating gas station hot dogs. I know how that goes.

The unloading on Silt Mesa went pretty smoothly. Erving slid down the ramp like a skier in his slippery shoes, while I steadied the hand truck from the side.

At Paul's when I wrote him a check, he said, "Thanks, and thanks for making my lunch."

Thanks? Thank you, Amigo. I'd have never got it done on my own.

November 2008

Lettuce Patch yard on the Flat Tops. © photograph Irene Owsley.

Dead man walking

Some days turn out unforgettable, in the oddest and most unpredictable ways!

A few Colorado Octobers ago – never mind how many – we got Paul's bees ready to go to California for the almonds.

We worked in pairs. One of us would tip the hive forward, while the other used a hive tool to pry off the pollen trap. Then the first would lift the hive, while the second removed the trap.

The Bookcliff Range was already white, and snow swirled around us as we worked.

The bees were all in the hive, at least until we riled them up. I don't recall that anybody had a smoker going. We just did it. Some hives scarcely stirred, but in others, the little darlings came out with a vengeance.

When I stepped away from the mayhem to relieve myself, I got stung not once, but twice. This really got my attention! I cried out, and Mark said, "Come on! How could any bee find such a small target?!"

The entire crew got a laugh at my expense. They were such a good-natured bunch.

By noon, we were cold and a little beat up by the bees. When Paul said, "I'll buy lunch in Battlement Mesa," nobody objected.

The White Buffalo was the only joint in town.

As we pulled into the parking lot, my old friend Jim Bare came out of the restaurant headed down the sidewalk toward City Market. I saw him clearly from 25 yards away. He walked slightly hunched the way he always did, with his hands in the pockets of his old red ski jacket. I almost jumped out of the truck to say hello. Looking back, I wonder why I didn't.

It didn't surprise me to see Bare, because I knew he hung out at the White Buffalo. We'd had lunch there awhile back. He was in poor health. I'd meant to call him again, but you know how it is.

After we all ordered, I eased over to the bar. The bartender smoked a cigarette as he wiped the bar top. A couple of aging longhairs sat staring into their beers. I wanted to leave my beekeeper business card with the bartender, for Bare when he stopped in.

"Say," I said to the bartender, "You wouldn't happen to know a guy by the name of Jim Bare, would you?"

"Can't say that I do," the bartender said. This set me back, because I knew Bare was a regular here.

I said, "He lives in Battlement Mesa. You'd just about have to know him." Then I described a certain unmistakable physical peculiarity of Bare's. It wasn't something I'd ordinarily mention, but just this one time I did.

"Oh, that guy," the bartender said, "He died last spring. We had a wake right here in the bar." The old hippies looked up and nodded knowingly. The bartender continued, "You know, he special ordered a bottle of his favorite single malt scotch. It's still here. We're waiting for him to come get it."

All this hit me like a stomach punch. Bare was a friend and an easy man to like. When I was a stonemason, he was my hod carrier. We somehow got into a wrestling match once. We took a road trip together. A lover of books, he had a certain scholarly air. A born storyteller, he spoke only cryptically about his novel – a work in progress. Maybe as a writer I felt a certain kinship.

Walking away from the bar, I had this eerie feeling. Back at our table, I said to Paul and the boys, "I don't get it. I just saw a guy in the parking lot, but the bartender says he's dead. Now who do I believe – a guy who doesn't even know my buddy's name – or my own eyes?"

There had been women in Bare's life, but I was pretty sure he lived alone. I knew he had a grown daughter. When I dialed his number, a woman's voice answered on the machine. I left a message for Bare. I said I'd heard a rumor that he might be dead. Would he mind calling and setting me straight? This was an odd message for sure, but if Bare were somehow still alive, he'd be most amused.

No one ever called me back.

Finally I called Bare's lifelong friend in Aspen, and he confirmed the worst. OK. I can accept this. I have to. But who was the man in the red jacket?

My Brazilian friend Dea is the daughter of Helena Vieira Costa – "Mother Helena" -- the legendary psychic and orphanage founder from the city of Alagoinhas. Even today, over a decade after her death, Mother Helena's fame as a medium and near-saint rests secure – not only in Alagoinhas, but to the farthest reaches of the state of Bahia.

All this being a way of saying Dea knows her way around the netherworld. Though no seer herself, for Dea the spirit world nonetheless forms a sort of backdrop to her earthly life.

Over lunch in El Jebel, Dea listened bemusedly to my story about Bare. Maybe her dark eyes twinkled. I said, "What do you make of the guy in the red jacket?"

She said, "That was your friend. He showed himself to you as a way to say hello."

I said, "You really think so?"

Dea laughed that anyone could be so skeptical. "Of course," she said. "Who else could it have been?"

March 2010

Granny

Last year Granny politely asked if I might put some bees on her property – for the garden. She wants more raspberries and bigger apples.

"Sure, we'll talk about it," I murmured, hoping she'd forget.

Granny wasn't offering space for a 30-colony bee yard, you understand. She wanted one hive out behind the shed or on the vacant lot next door that she doesn't own -- to pollinate her fruit and garden blossoms.

Granny lives in Carbondale, but I don't. And while I keep some bees near Aspen for a few weeks in the spring, after that I hardly ever get up that way. Honeybees require care and supervision. I wasn't about to drive 45 miles every two weeks to tend a solitary beehive.

A couple of months ago Granny brought it up again. I wiggled. Then last week she called. She turned on the charm, like she does when she wants something.

I sensed an obligation. After all, I passed many a summer evening fishing at her place. I never turned down a meal when I dropped in unannounced. Plus Granny and I do go back some.

Long ago she ran the Snow Chase Lodge, a ski bum flophouse at the base of Aspen Mountain. I stayed in the bunkhouse out back. Then she rented me kitchen and bathroom privileges when I moved into a nearby abandoned ski-lift shack.

When my shack's coal stove blew up in my face, and I thought I might go blind, she drove me to the hospital.

When I shot a four-point buck, she instructed me to hang it by the cellar door and taught me to butcher.

When my dog trotted over Trail Rider Pass and turned up in Marble, Granny picked her up.

She claims she taught me to ski. Actually, we did ski a day together at Buttermilk once, and she did bark out "tips."

When she moved from Aspen to a house on Highway 82 outside of Carbondale, the deer carnage on the road bothered her. For a time, when she'd hear the screaming tires and telltale thud in the night, she'd go forth in her nightie with her Winchester .30-30. Stunned drivers and sometimes entire families watched in shell-shocked disbelief as she dispensed her mercy and then dragged Bambi down to the garage.

At some point she changed her name from "Mrs. Mac" -- never her real name -- to "Granny." Only complete strangers and her very closest friends call her Pam.

Her diet centers around Coca-Cola, a particular brand of sherry, and meat. She harbors strong opinions about what constitutes a healthy lifestyle. She holds strong opinions on just about everything.

Granny hops around in the garden, cussing and setting traps for "varmints." She straightens her ball cap and squints into the sun, then takes a hard hit off her unfiltered Pall Mall.

"There aren't any damned bees around her anymore. That apple tree used to hum with bees at blossom time," she says.

I knew I owed Granny, but I didn't want to pay up. I decided to pass the buck.

When I called Kay in Carbondale on a Sunday morning, she was reading in bed. She said she was down to three hives. I own 80, and I was about to run down to Collbran to check on some.

Kay said she'd never met Granny, but she found Granny's property location intriguing. Plus, she said, "I like old people." I spared her some details. She'll learn soon enough. There won't be any helping Granny out of bed or keeping her pills straight. Kay might get to help her dispatch a pesky skunk, though. Just please don't step on Granny's snakes. She really does love her snakes.

Kay thought she ought to move a hive over to Granny's soon -- before it got too heavy with honey. I liked the sound of this.

OK, Kay didn't actually commit, but she did take Granny's number.

When I called Granny and told her all this, she chirped about more and better fruit and veggies and said she wouldn't wait for Kay to call. Granny said she'd call Kay right away.

Like a prayer answered, later that day I got a swarm call from Carbondale. This gave me an excuse to bug Kay again. "Look," I said, "You could hive this swarm and drop it off at Granny's. Save you from moving one of your heavy hives."

Kay hesitated, but she didn't say "no." She could have sounded more enthusiastic, but maybe she was still reading that good book in bed. "I haven't had a lot of luck with swarms," she said. "It seems like they die out or just leave the hive."

I tried to buoy her. I said my luck was the opposite. I said swarms provide a free way to increase your holdings. Mine generally proved gentle and productive. Kay sounded skeptical but said she'd phone the swarm caller.

One way or another, Kay's going to take Granny some bees. I know she will. She has to.

Otherwise it's up to me.

September 2003

Granny . Photograph courtesy of Tracy Delli Quadri.

The world's most famous beekeeper

I never dreamed I'd meet the world's most famous beekeeper, or that he would be so likeable.

It happened this summer at the first of my weekly bee seminars on Aspen Mountain. I work as a "ranger" on the ski hill, and my boss thought a bee talk might be an interesting diversion for the tourists who ride the gondola to the top of the mountain for lunch, a bluegrass concert, or an alpine hike. It was a gamble. Who knew how popular this might turn out to be?

I was scheduled to begin my talk at 11 a.m., but by twenty past, no one had showed up. I was about to give it up when a family of four walked into the ski patrol shack --two teenage girls, an attractive woman of 45, her husband perhaps 10 years her senior. The woman approached me and said, "We're here for the bee seminar."

Eureka! I had a class! "You're in luck!" I said. "I'm your teacher, and you're my only students!"

"Excellent," she said. "My husband does not speak English, so I will translate for him. He is the president of the Brotherhood of the Ukrainian Beekeepers."

The few hairs left on my bald head tingled. I locked eyes with the man. I instantly recognized him. There was no doubt.

"Your husband was also the president of Ukraine," I said.

"Yes," she said. "He was."

He smiled at me and extended his hand.

And so it was that I met Victor Yushchenko. You already know all about him. Everybody does. May I refresh your memory?

The year was 2004, and beekeeper Yushchenko was running for president of Ukraine, when events took a most bizarre turn. Someone poisoned him. He survived, but his face became grotesquely transformed from "chloracne," or chemically induced acne. The press dutifully sent his caricature of a mug shot 'round the world.

Ukraine -- and the world -- was electrified. Rumors swirled. Who could have done such a dastardly deed? Could it have been his principal opponent, Russian-backed Viktor Yanukovych? Might it have been the KGB?

The toxicologists reported that his body contained massive amounts of TCCD, a potent form of dioxin – a key ingredient in the defoliant Agent Orange, used by American forces in the Vietnam War.

A run-off election took place in November, with Victor Yushchenko facing Viktor Yanukovych. The beekeeper Yushchenko led by a wide margin in exit polling but Yanukovych was declared the winner, amid widespread rumors of vote-rigging and voter intimidation.

Millions of Ukrainians took to the street in a series of non-violent sit-ins, protests and strikes against the election results, ushering in the famous "Orange Revolution." The Ukrainian Supreme Court ordered a re-vote, and under the watchful eyes of observers both international and Ukrainian, beekeeper Yushchenko defeated Yanukovych on the day after Christmas, 2004, to assume the presidency of his country. He served until 2010.

I'm no expert on Ukrainian politics or even honeybees, really. What does a hack American sideline beekeeper tell the president of the Brotherhood of the Ukrainian Beekeepers about bees?

The correct answer is "not much." I explained how my little 80-colony operation works. I told him about important local honey flowers. I told him about my ten-hive bee yard almost under the gondola near the bottom of the ski hill. We talked about nosema and Varroa. I told him about our Colorado bears.

That took an hour. He asked a lot of questions. Then he turned the tables and told me about beekeeping in Ukraine, the fifth-largest honey producing nation in the world.

Beekeeping has been an important part of Ukrainian culture and commerce as long as anyone can remember, Yushchenko explained. In 1814 the Ukrainian Petro Prokopovych invented the first hive bodies with removable frames. Prior to this, Ukrainians marked "bee trees" to establish ownership, sometimes cutting down the trees to leave "bee stumps."

While large-scale "industrial" Ukrainian beekeeping operations employ the familiar Langstroth hive system, traditional backyard and sideline beekeepers are more apt to stick with the "Ukrainian" or "Slavic" system, which uses 24 very large (29 X 43 cm.) frames. If I understood Mr. President correctly, this hive is not honey-supered. He feels this way of keeping bees is less disruptive to the little darlings, as it avoids squashing them as I often do when I super my Langstroth hives.

On a sheet of paper he illustrated Ukrainian hive types as he talked. "My husband is very artistic," Mrs. Yushchenko said.

He said the favored bee in his home country is the Georgian, noted for its calm demeanor. He waxed rhapsodic about honeybees in general, which he calls "God's insects." He said we should "respect and love," not kill them. Gentle Yushchenko clearly has a love not only of bees, but of his country, and its traditional culture in which the honeybee prominently figures.

Finally he said, "Apimondia 2013 is in Kiev. I'm the host. You should come."

Me? Fly across the Atlantic Ocean practically to Russia for the biggest bee meeting in the world? Ever the thrifty traveler, I said, "How does the dollar fare in your country?"

"You'll be fine with your dollars," he said. "There will be a week of Apimondia and then a week of tours. You should come."

His wife Kateryna blurted, "I'll give you my personal e-mail. Write to me. I'll find you a place to sleep."

To say I was touched would be an understatement.

After we talked ourselves out, he kept shaking my hand. Odd the way those little darling honeybees can bind strangers together, no?

Later, Kateryna wrote to me: "Our kids and I wanted to go to the top of Aspen Mountain, my husband did not. I went to the kiosk to buy tickets, and saw the sign about your lecture. I told him, and he immediately agreed – it was like an invitation to him alone!"

I walked the Yushchenkos to the door. I offered a lunch recommendation. I told them to be on the lookout for my little apiary under the gondola. He shook my hand again.

I don't always think on my feet. We never posed for a photo. I never gave him a ski patrol ball cap. It never occurred to me to invite him to take a peek inside my Aspen Mountain hives.

Ukraine . . . not a place I thought I might ever visit. And yet . . . a personal invitation from the Apimondia host and world's most famous beekeeper . . .

Think of it -- *Apimondia* – the word rolls off your tongue. Imagine the sights, the comradeship, the friends a beekeeper might make in Kiev.

I'm old now. Life's too short. Maybe I should do this.

October 2011

The passing of Granny

Six weeks ago, as I was leaving the familiar clutter of her cigarette-scented kitchen, Granny followed me to the door. "You should stop by more often," she said, with the sweetest sincerity.

She was a people person. She knew I drove past her house every Sunday evening, and she loved nothing more than company. So I promised, and I meant it, but I never saw her again.

She was only 85. People would ask me how she was doing, and I'd say, "She's great! She looks 100, but she only acts 50!"

She definitely didn't wear those orthopedic granny shoes or use a walker or spend any time getting her hair done. She chopped off her waist-long braid years ago, but she still wore the same work boots or sneakers, Levis, ragged denim jacket and U.S. Ski Team ball cap.

For years Granny didn't have much good to say about honeybees. When she was a child, she "nearly died" from a bee sting. She was "violently allergic" to bee venom. She'd go on and on if you let her.

Then, in her golden years, she retired to the garden, and guess what! She decided that she wanted bees to pollinate her beloved raspberries and her squash and her apples.

So would I mind leaving a hive or two behind her shed, on Barbara's vacant lot next door? It would be wonderful for my bees if they could visit her cucumber blossoms, wouldn't it?

To be honest, Granny never was my granny, but we did go back to when the world was young. She was a mentor and an icon – an irreverent, mountain-climbing, deer-hunting, powder skiing Aspen ski lodge owner who would rent you a bed for two bucks, or a buck-fifty if you could tolerate the bunkhouse. Her ski bum tenants called her "Mrs. Mac" or simply "the manager." She was like a mother, but she was oh, so cool. I adored her. I still do.

So now, 40 years later, even if putting bees at her place seemed an inconvenience, how could I say no?

Granny's place of retirement is across the river from the edge of Carbondale, in a patchwork of small subdivisions and horse properties. There's some rabbit brush but not a lot of alfalfa or sweet clover. I decided that bees would never thrive here, and for a time I made this a self-fulfilling prophesy.

I wasn't doing this for me. I was doing it for Granny.

The first summer, when I dropped off a couple of weak-sister hives, Granny said offhandedly, "There's a bear that comes around at night."

Oh, great -- a poor location, with a bear. I erected a solar electric fence, and then I said to Granny, "Do you have any bacon? I need to drape some on my bear fence, so when the bear comes sniffing, he'll tickle his nose."

"I can get some," Granny said enthusiastically. "I can take care of this."

"Are you sure?" I said. "You're allergic, remember?"

"Don't worry about me," she said with a devilish grin.

My bees made no surplus that summer, but those pathetic little darlings wouldn't have made honey in the Garden of Eden. Yet Granny was ecstatic about her improved yield in the garden.

The following year I again brought two colonies, and right away one went queen-less. When I caught a little swarm in town, I united it with the queen-less hive. The result was a two-super honey surplus for that hive, mostly on late summer rabbit brush.

Hey! I was doing this for Granny, but maybe this wasn't such a bad spot. After a few more good seasons, last summer I dropped off eight strong hives, and they were my overall best producers. The honey was alfalfa-light and mild, not dark and butterscotchy like rabbit brush. Go figure.

Last Wednesday morning I was thinking, "This year I'd better get a pickup load of bees to Granny's in time for the dandelions."

When I went into the house, there was a message from Monk. "I have some bad news," he said, and my heart sank.

Of course. Heart attack or stroke. It had to be. You don't smoke a million-plus unfiltered Pall Malls, live to be 85, and then cheat the doctors.

Imagine my consternation when I learned that Granny had been run over right in front of her house!

She was returning from a visit to the home of her best friend Barbara, who lives directly across Highway 82 from Granny. At 10 p.m., while crossing the road on her way home, Granny was struck by two cars and killed instantly.

Understand that Granny and Barbara were friends for 40 years. They lived 100 yards apart, and they visited every day. Barbara always drove across Highway 82 to Granny's, and when Granny visited Barbara, she always walked. That's the way they did it for a quarter-century.

As for what actually happened, we'll never know. Granny may have had difficulty judging the distance of oncoming cars. She had some recent history of losing her balance. She might have fallen.

Traffic on four-lane Highway 82 can be horrendous. Barbara said Granny sometimes made it halfway across before waiting in the middle turn lane for a break in the last two lanes.

What? Oh, Granny, how could you!?

In my mind's eye, I see the surprise on her face in the headlights. I feel the horrible thud. Her ball cap flies off, as her poor frail granny body floats through the night, her little arms and legs flying every which way, until finally she lands like some rag doll, clear down by Barbara's vacant lot, down where I keep the bees.

May 2009

Slim and the Dalai Lama

By 2:30 a.m. on April 1, it had already been a tough week. First, Slim gave me my pink slip. She had her reasons, don't you know, but that didn't make it any easier on this old fool. I bravely told her, "You gotta do what you gotta do, but thanks for the memories, you little darling!"

But this time it was bees, not a broken heart, that woke me in the night. A tied-down load of 'em on that one-ton flatbed in the driveway got me tossing.

Every spring I haul honeybees to Grand Junction and Palisade to pollinate the orchards - first apricots, then sweet cherries, then pears and J.H. Hale peaches. That's three loads 150 miles roundtrip from my place in Peach Valley, mostly on I-70. You could call it a risky trip on good roads in a bad truck.

The Fords I drive are both notoriously unreliable. They just are. They're ancient. The newer one is over 20 years old. Don't ask me why I don't invest in a decent rig. I'm just too cheap.

I like to take the pickup, as opposed to the flatbed, because it's lower and easier to load. I put a ramp up to the bumper and pull a hand truck.

But once I got the pickup loaded with 11 hives, it wouldn't start. This wasn't the first time the beast had let me down, but every time I repair it, I think, "The engine's sound. I don't drive that many miles. Investing in a newer truck is crazy, considering what I'd have to pay." But it's always some little relay or computerized component – not the engine -- that fails. I never learn.

In the twilight I moved my bees from the pickup onto the flatbed, but that old wreck has its problems, too. It's a hard starter, and that's what got me up that brutally cold spring night. I'd just gotten that truck back from the shop and hadn't had a chance to see if the cure took. It was 15 degrees outside, and I wanted to be sure I could get it going. When I turned the key, it protested mightily, then finally coughed and sputtered. Five minutes later it purred.

I'd ordered a bee net, but it hadn't arrived. The prospect of breaking down on the Interstate with an uncovered load of bees was not a happy thought. I had no backup truck and no cell phone. I could picture the highway patrol turning a bee truck breakdown into a full-blown crisis.

I was on the road by 4:30. You don't expect to deliver bees when it's 15 outside, and only 70 miles away, the apricot buds have already broken. But that's Colorado for you. It was so cold I could have departed at a civilized hour. The bees would have never stirred until late in the morning, if then. But I'd agreed to meet a new customer at 7, after I dropped off my first two loads. The cold snap caught me by surprise, and I wasn't about to call my guy in the middle of the night to re-schedule.

I actually like to be on the road in the wee hours. The traffic was all headed the other way, making the commute from Rifle, Debeque and Grand Junction to the natural gas wells and the money towns of Aspen and Vail. Some folks spend three and four hours a day on the road. I can't imagine it.

Natural gas rigs lit up the whole Colorado River Valley. You don't comprehend the magnitude of the boom until you see it at night. Lights stretched from Battlement Mesa clear up to the Roan Plateau.

I thought about Slim leaving, and I can honestly say I was at peace with it. Not to be maudlin, but I can see the rainbow's end. I told myself (and I'll tell you!) that I'm on the Ten-Year Plan. I'll play the cards I'm dealt, but if the Good Lord goes along with the Plan and gives me ten more years of health and vigor, I'll have drunk a full cup. There won't be any grandchildren bouncing on my knee in the last days, but perhaps I might still decline with some measure of grace.

I wasn't afraid to explain this to Slim. I'm too old to be shy. There isn't time. I told her she could be part of the Plan. Or not.

The stars were only beginning to fade as I rolled into Grand Junction. The thermometer read 23 at Roger's place.

A retired physician and a kind and learned man, Roger likes sweet cherries, and bees. He wishes I'd leave mine all summer. A devout Christian, he recently visited India, where he had an audience with the Dalai Lama. "The Dalai Lama has an interesting perspective," Roger said carefully.

His half-acre of sweet cherries lies completely surrounded by the city. Town grew up around this oasis. It unnerves me leaving bees there, because the neighbors are too close, but the good doctor is so appreciative. I hid the little darlings behind a bush.

I made the rendezvous with my new customer and dropped off four hives in a cherry orchard set against stunning cliffs by the Colorado National Monument. The grower said he forgot his checkbook.

Now the sun was up. At breakfast in Palisade, I kidded the waitress a little. She called me "Honey." On the road again, the world looked right.

June 2008

The check's in the mail

My beekeeping days almost ended December 3 when I got buried in an avalanche ski patrolling on Aspen Mountain. Maybe I even caught a glimpse of Heaven.

I was blue (some say purple) when they dug me out. The first face I saw, the first voice I heard, was Ali's.

"Ed, we love you," she said over and over. She sounded so desperately sincere. Everybody who was there says I smiled.

My adventure became the talk of the town. I got phone calls. I got hugged by women and sometimes even men I barely knew. None of this short-lived celebrity hurt my honey sales one little bit.

I sell some honey right out of the ski patrol headquarters at 11,212 feet. Not a lot, but a few hardcore skiers do go down the hill with 3-pound plastic jugs of honey tucked inside their jackets.

I use a low key "don't ask, don't tell" marketing strategy, and so far management has rewarded me with a wink and a nod. My only advertising is my beekeeper's card on the patrol room door, with "Inquire Within" hand-printed on it. But word gets around.

The other morning I filled my backpack with 12 quarts and skied down to Bonnie's Restaurant halfway down the mountain. It was a Saturday, so I knew there'd be not only the Latino kitchen crew to sell to, but also the Swiss pastry ladies.

When I arrived, Manuel and Santos were shoveling the deck. Felipe was shoveling the roof. They asked, "How much?" When I said "Nine bucks, just like last year," Santos immediately reached into his pocket. Felipe leaped off the roof into a snow bank and said, "How many do you have? I'm sure Hugo wants one, too."

The Latinos around here are crazy about honey. But they always look me in the eye and ask me if it's "pure." Back home the custom is to cut honey with some other sweetener, and they seem incredulous that I sell unadulterated product.

Felipe handed me a ten-dollar bill and refused to take his dollar change. You gotta love these guys.

Inside the restaurant, the word was out. The Latinos were all buying. The Swiss pastry ladies pushed some and poked with their sharp elbows. It was a buying frenzy, complete with a touch of panic that I might run out. Within five minutes, I'd cleared $108.

The owner took the last jug. She might be 35. Her father ran Aspen Mountain's Sundeck

Restaurant for 28 years. The whole family lived on top of the mountain. She skied to school, just like in some magical children's story.

The ski patrol director from another ski area called the other day about eight-ounce honey bears. At first he said he wanted 400. I said, "Whatever for?"

He said, "For ski safety week. We're giving them away. We'll get you some free advertising. I'll need them Saturday."

I said, "I see. Instead of pulling their tickets, you'll stop the speeders and say, 'Slow down, Honey,' and give them a bear, right?"

"Not exactly," he said. But we're calling it 'Colby Farms Slow Zone honey.'"

I didn't get it, but the director did, and that was what mattered. I told him I'd come up with a price.

I decided $1,000 would make it worth my while. I'd have to rush-order the empty bears, and melting and packing honey would occupy both of my days off. And I wasn't going to compete with Wal-Mart.

When I called back, he said, "$500 is all I have in my budget."

I said, "OK, I'll give you 200 bears for $500."

He said, "How much are those bigger honey bears you sell to the restaurants?"

I said, "Three dollars."

He said, "Three dollars? That's all? Three bucks? I'll tell you what. I'll take 50. That would be $150, but I'll give you $200. Would that be worth your while? And next year we'll do a big order."

OK, I wasn't going to make a killing, but all of a sudden this project looked manageable. And I don't have a lot of spare time in the winter. I liked the idea of a $50 tip. I said sure. I said I'd drop them off.

That night in our patrol locker room, I told this story to Bud, who is a contractor when it isn't snowing, and who knows about deals. He said, "What do you think about that extra 50 bucks?"

"I like that extra 50 bucks," I said.

Bud said, "When a guy makes an offer like that, I think it's always smart to give him what he pays for. That way he'll think you did him a favor, and not vice-versa. That's just my opinion, Ed."

I was going to ignore Bud's advice, but on reflection I decided to deliver 67, not 50, bears for the $200.

On delivery day, the patrol director called me at home just as I was about to head out the door. He said, "I know I promised to leave you a check today, but I found out I need your invoice first."

"No problem," I said. "I know you're good for this."

When I dropped off the honey, I made sure I put "67 honey bears" on the invoice.

I think Bud's advice was good advice. Now I feel like I have the upper hand. It's been a month now, but I'm not worried. I know they're good for this. I'm sure the check's in the mail.

April 2006

This is Cuba

The driver of our bombed-out 1956 Pontiac taxi was going much too fast. He seemed to be having trouble staying in his lane. Where's a seat belt when you really need one?

My sidekick Marilyn and I sat in the front seat. She put her arm around me, smiled and said, "Life is good."

I looked down the eight-line highway. Traffic was light, with horse-drawn buggies outnumbering cars and trucks. Graceful royal palms dotted the verdant countryside. In the far distance, a sugar mill belched thick black smoke. Alongside the road, a doctor and a nurse hitchhiked in their hospital smocks.

"This is Cuba," I said.

When we turned onto the two-lane road to Vinales, we passed by a cluster of houses. Up ahead, I saw one bicycle on the left, one on the right, and some pedestrians. The driver blew his horn and accelerated.

The bicyclist on the right wobbled across the road in front of us. Our driver swerved. There was a sickening thud.

At first I thought the driver was going to keep on going, but then he abruptly brought us to a stop. As I reached for the passenger side door handle, he put his arm across my chest. "Be calm," he said

Be calm? For all I knew, that was a corpse lying on the road back there.

The driver got out and did a quick front bumper inspection. Then he walked back to the crash scene.

By this time, the bicyclist was on his feet holding his bike. Its front wheel and front forks were mangled. A bystander pointed at the bicyclist and said, "He's drunk!"

First the bicyclist and the taxi driver shook hands. The bicyclist said he was fine. Within two minutes we were on our way again. No money exchanged hands. No one filled out any paperwork or called a cop. This struck me as odd, because after all, this is a police state.

We were on our way to Vinales to find some Cuban bees, and at the private home where we rented a room, our gracious hostess informed us that the chicken farmer next door knew some beekeepers.

I knocked on his door. Mayito said sure, he'd set up an introduction with the beekeepers. He recommended for transportation an illegal taxi owned by his friend Antonio.

The car was a mint condition, jet black '47 Chevy with the original straight-six engine. It was Antonio's pride and joy, not to mention his sole means of livelihood.

And he was as careful driving as our earlier driver had been reckless.

It was 15 or 20 miles to the bees, and when we got there, the beekeepers were somewhere else. But an affable apprentice offered to show me around.

First we looked at stingless bees that lived in a two-and-a-half-foot-long, oblong wooden box above someone's doorway. They looked pretty much like ordinary honeybees. They stored their faintly bitter-tasting honey not in combs, but in spacious, waxy mosque-shaped reservoirs adjacent to a brood chamber made up of horizontal sheets of comb similar to the comb of our own apis mellifera. These bees are apparently not raised commercially.

It was pleasant to look at bees without the obstruction of a veil, and I asked about Africanized honeybees in Cuba. Julio said, "We don't have them."

"Yet," I replied.

We looked through some apis mellifera hives set out neatly in a row in the shade, resting on two long pipes suspended off the ground. They were just getting on a honey flow.

This 900-plus hive migratory operation sells its honey to the government. They have a coastal season and a mountain season. They burn their American Foulbrood hives and practice drone brood removal for varroa. Terramycin is illegal. Julio said they never use it.

He'd never heard of Nosema.

He was such a good natured kid, and he knew a lot. When our visit was over, I tried to give him some money. Unless they have relatives abroad, pretty much everyone in Cuba is desperately poor, even professionals like doctors and engineers. Culturally, they're middle class, most of them. They dress pretty well, They read and write. They have basic health care. But they haven't got two nickels to rub together.

So when Julio refused my offer of monetary compensation for a service I requested, I didn't know what to do. Ten or 20 bucks would have been a small fortune to him, but I didn't want to offend him. He said he was just happy to be of help. Maybe I got a little misty-eyed when he said that.

I'd brought bee gloves that I thought I might give away, but later I reflected that they were caked with American propolis and wax. I didn't want to introduce some American virus or bacteria to the island of Cuba. So I left them in my suitcase.

I did bring a brand new Mann Lake "Pollinator" jacket with zip-in veil. Not cheap, but a quality garment made in USA. I rush-ordered it especially for this trip. I wanted insurance for dealing with possibly Africanized Cuban bees.

Our last morning in Havana before we flew out, I noticed the jacket was gone. I don't know where I might have left it. It's lost. Somebody in Cuba will surely use it, but it almost certainly won't be a beekeeper.

If I'd have insisted, maybe Julio would have taken it. I never planned to give it away, but now I wish I'd made him take it. I really do.

March 2011

A Christmas story

The best time to sell honey is Christmas. In the Aspen Mountain ski patrol locker room, I harangue my co-workers, reminding them that my classy glass hex jars of high-altitude wildflower honey are the cheapest made-in-USA stocking stuffers they're going to find. They even have labels! And what man, woman or child doesn't just love honey?

Still, there'd been some grumbling about price. Gorp's one of my best patrol customers. He chided me that my annual price hikes exceeded increases in the consumer price index. "I remember when you were selling quarts for nine bucks!" Jimmy chimed in. My increases reflect the sky-high, nearly doubled wholesale value of honey, my sense of the local competition, along with, of course, my bottom line. I don't want to gouge, but why should I play the fool?

I'm in this sideline business, hobby, whatever you want to call it, not to get rich, but because it's in my blood. It has unique challenges. None of my customers has ever awakened in the night worrying about Varroa mites or American Foulbrood.

I like to keep it simple. My response to the whiners is almost always the same. "Cheapest honey in the valley." This might even be true.

At Christmas, I like to hit the ski school locker room, too. These guys are getting their first paychecks, along with fat tips from all the movie stars and Wall Street bankers. Honey flies out the door.

Last year I got behind, and the patrol cleaned me out of locker room honey. They wanted more. Meanwhile, my gal Marilyn was selling honey faster than I could de-granulate it. I wanted to make a special trip to Aspen on Monday – Christmas Eve – to peddle to the ski instructors, but I was running out of liquid honey. I figured I could de-granulate a couple of five-gallon buckets the night before. Who needs sleep, anyway?

There was a catch. I'd promised to go caroling with some patrol chums on the night before Christmas Eve. I didn't see how I could weasel out. The best I could hope for was that the caroling somehow fell through.

For some cynics, anything that smacks of religion is not so cool. Plus, unlike drinking, Christmas caroling is not a traditional ski patrol custom. And you have to sing, whether you hit the notes or not! But my pal Steve has a certain infectious enthusiasm, and this was his baby. He got a few takers. I didn't have the heart to tell him no.

On the appointed night, a handful of patrollers sat around the locker room after work, beer mugs in hand, waiting for caroling reinforcements from the other three Aspen-area ski patrols. "They'll be here," Steve kept reassuring us.

An hour later, we knew we'd been stood up. I was thinking about a long drive, and warming that honey. I figured I'd break up this party. I said, "If we're not going to carol, I need to get home to two lovin' arms!"

Our boss Bud leaped to his feet. "Thanks for throwing down the gauntlet, Ed!" he cried. "Let's do it."

My heart sank, but I jumped into the back of Bud's ancient pickup with the rest of the crew. We started off with a rousing chorus of "Jingle Bells" as we headed downtown. This wasn't so bad!

Aspen was packed for the holidays. People waved as we put our poor best into Silent Night, Deck the Halls, and We Three Kings. Children stared transfixed, as if they'd just seen the Blessed Virgin Mary or even Santa Claus!

Outside an eatery where dozens of paparazzi waited for a shot of somebody famous, all heads turned as we passed. I expected somebody to yell at us to take some voice lessons, but everybody on the streets acted as if they appreciated our willingness to get into the spirit, even if we really weren't very good.

I was freezing in the back of that truck and hoping we'd done our due, when Bud announced we were headed for the senior center on the outskirts of town. We arrived unannounced, but the person in charge agreed to ask the residents, most of whom had retired for the evening, if they might like to see and hear the ski patrol carolers.

Yes, indeed, they did want to hear some caroling! Seniors filed in on unsteady feet, with walkers and wheelchairs, beaming as they joined in on old holiday favorites.

We finally sang ourselves out. I never in my life felt so popular.

I got home late. I didn't get the honey warmed that night, or make it to the ski school locker room to cash in on Christmas Eve.

But maybe I'm the richer for it.

January 2014

A rude awakening

You never expect to see bears in the winter, but in December the ski patrol was doing avalanche control work on Aspen Mountain. Art and I taped a five-pound cast primer explosive onto a bamboo pole that we stuck into the snow, so that the charge was suspended in mid-air. This is called an "air blast," and the idea is to send shock waves into the snow and precipitate avalanches. Air blasts are loud. Art and I retreated a safe distance, cupped our hands over our ears, and waited.

The detonation reverberated through Rayburn's and the Cone Dumps. Unfortunately, and unbeknownst to us, the charge went off right next to a bear den. A bear came flying out and headed 75 yards down the hill before she stopped. We were just below a ridge. Other avalanche control teams were spread out across the hillside. As the other teams' bombs went off, our bear stood on her hind legs and clawed the air. She stayed in a clump of trees adjacent to the open slope below us, which is where we had planned to go next.

I was more than happy to change plans, however. Artie concurred. "No way I'm going down there," he said.

To go back up would mean a steep slog on skis in deep snow. Our route to the left was closed off by trees and cliffs. To the right would take us just below the bear den and directly above the bear, but after that we could make a long traverse across the hill. I was anxious to move.

I said to Artie, "Let's get out of here," and headed to the right.

Just as I set out, another patroller called out on the patrol radio, "There's another bear!" I looked uphill in time to see a cub wiggling his ears and looking down at me from the opening to his den 20 feet above. I could just see his head. The little darling was making high-pitched grunts that sounded a lot like "Mama, Mama!"

At this point, I was directly between the enraged mother bear and her frantic cub. Snow conditions were not ideal. In September we had a heavy snow that knocked over fully-leafed aspen trees. Now I was skiing over rocks and downed trees in deep snow. I'm not saying I panicked, and I'm not saying I didn't. I moved with haste and was careful not to fall.

We got out OK. I took some ribbing over the radio, to the tune of, "Colby, where's a pail of honey when you need one?" But we all lived to hoist another locker room beer.

The den was in a place that's hard to get to, but J.T. and I went back later and fenced it off to keep out skiers and remind patrollers, so that maybe our furry friends could slumber yet.

Around here, most bears respect an electric fence around a bee yard, most of the time. But with 85 Colorado bee yards, Paul occasionally has a "problem bear" out there somewhere –

a bear with a taste for honey and larvae, who'll rip down the stoutest, hottest solar-electric fence to get some.

Colorado bear numbers have been on the rise ever since the 1992 passage of a controversial referendum banning the traditional spring bear hunting season. Hunters felt betrayed. The enviros figured they won one. Nobody ever thought about the beekeeper.

Paul claims that bruins have to learn about the banquet of delicacies within a bee hive. Initially a bear might merely be curious. That's why the first time one visits an apiary, you might only see tracks. He might knock a cover off. The second visit he might tip over a hive. It's maybe the third or fourth time that he discovers the delights of raking honey and brood into a gluttonous maw.

Paul's wife Nanci runs a wildlife rehabilitation center. She loves nothing more than rescuing scrawny orphaned bear cubs, fattening them up through the fall, then singing them lullabies until they dream white winter dreams. Along about February she sedates the little darlings and gets the Division of Wildlife boys to help her haul them to cozy little dens that she knows about. She does this every winter, and she claims she's never lost a bear. Without Nanci's help, virtually all of these little critters would have succumbed to the cold law of natural selection.

So you can see why Paul sometimes has bear problems.

Paul is the president of the Colorado State Beekeepers' Association, and we've had our meeting at his and Nanci's place the past two summers. You couldn't ask for a sweeter, shadier spot to spit watermelon seeds and talk about apis mellifera. Nanci happily offers attendees a tour of her wildlife center, which is right there.

Personally, I find it charming beyond charming that a man and a woman might pursue disparate dreams and yet find common ground. We should all be so lucky.

I told Nanci about the bears and the bomb. She laughed and said, "I read something about that in the paper."

When I teased her about being part of the "bear problem," she said, "I'm not responsible for any bad bears. My bears are all ear-tagged, and they've never bothered bees."

I said, "How would you know?"

She said, "Well, the rancher always knows about the bear that's around, and those bears never have ear tags. Those problem bears are big old bears. My bears are cubs."

I said, "So your bears never get old and raise hell?"

"Nope," she insisted. "My bears never cause trouble."

April 2007

The bear took a bite. Marilyn Gleason photograph.

A levitation

The registration line at Apimondia, the international bee conference in Kiev, Ukraine, took five hours, mostly outside in the rain. Once inside the building, it got tight. The crowd hydraulics were so powerful, our friend Therese levitated and tilted sideways in the crush. Think of it! People couldn't breathe. Some cried out for mercy. Four registration booths handled thousands of conferees. One copy machine and not enough toner. The Germans nearly rioted. No one died.

That was on Sunday. By Monday, when my sidekick Marilyn and I registered, it was a relative breeze. At least it wasn't raining. With one of us to hold the other's place in line, we drank espresso, ate open-faced lox sandwiches and learned how to say "toilet" in Russian. We struggled to communicate with cheerful Ukrainian beekeepers, threw our arms around strangers and posed for pictures. An English-speaking economist with shocking red lips, in a sleeveless black fur dress, invited us to visit her in Belarus.

There were odd moments, like when we finally arrived at the door to get inside and met a horde of already registered conferees being denied admittance to the building. They wanted to get to a lecture upstairs. The official language of the conference was supposed to be English, but all the door guards could say was "Nyet!"

Two hours later Marilyn and I had our conference passes, and we could look back and laugh at it all. Airline travel has trained us to accept and even expect the absurd. You learn to shut up and wait in line. But Therese's levitation story riveted this claustrophobe. I wonder if I'd snap!

Inside the hall, the tradeshow featured hundreds of exhibitors. I learned about high-tech Lyson Polish honey extractors, Austrian BeeVital herbal chalk brood treatments, a Romanian machine to extract beebread, the expansion of beekeeping in Tanzania.

The lectures ran the gamut, from dynamic to inscrutable. We native English speakers get spoiled. We expect the world to come to us, and it does. But if you attend a talk titled, "East Java Propolis Inhibits Cytokine Pro-Inflammatory In Odontoblast Like Cells Human Pulp"or "The Swarming Industry: study of the unique structure of Ukraine's beekeeping branch, it's origin and stagnant stability," be forewarned: you may be in over your head.

The University of Maryland's Dennis vanEngelsdorp hammered on a favorite theme: the inability of some beekeepers to adapt to change. He showed a bell curve that graphed people according to their willingness to accept new ideas. Then he polled his audience about their cell phone use - who got a cell phone early on, who waited a while, who had an I-phone, and finally, who today still doesn't have a cell phone. That would be me. Marilyn and I were in the front row, and I was too sheepish to look back and see who else raised their hand.

We went on a three-day Ukrainian beekeeping tour with other Apimondia visitors before the conference, so luckily we had a cadre of new friends. Nothing like a familiar face halfway around the world!

I was eating at the sandwich wagon outside the main hall when I spied a spare, graying gent letting his freak flag fly. He sported a blue jean jacket that advertised "CC Pollen, High Desert Pollen, Phoenix, Arizona." I thought, "I wonder who that is. Must have worked for Bruce at CC Pollen once."

Then it hit me: That had to be Bruce! I'd only heard his gravelly voice on the phone before, and in my mind's eye, he'd looked a bit more, oh, businesslike, maybe. That phone voice for sure had short hair. I wolfed my sandwich and ran after him, but he'd already vanished into the crowd. Later, we caught up, so I got so say, "Bruce Brown, I presume?"

Bruce is a smart guy and an international operator, but he doesn't show you all his cards right away. We kept running into each other, finally outside the catacombs of the Kyevo-Pecherska Monastery, where a beggar approached us. Using unmistakable sign language, the beggar informed us he was hungry. What can you do? Jesus commands us to feed the hungry. We gave him some hryvnias. But it didn't end there. The beggar demanded more. It got ugly. A shopkeeper with a stick finally chased him off. Tourists, we're always a mark.

Later, Marilyn got her pocket picked in the subway. The thief unzipped the pouch hung around her neck. He never found her passport or most of her cash, tucked inside separate compartments. She lost her debit card and her Garfield County library card. I said, "Some pickpocket's going to come to the States and check out a bunch of books on your card, and you'll get the fine when they don't come back!"

So we had a little bad luck. We had good luck in Ukraine, too. I'll tell you all about the good luck. Another time.

December 2013

The beekeeper handshake

Our hosts promised that a driver would meet us when we landed in Kiev, Ukraine. With a formidable language barrier, I wondered how we would connect.

Not to worry! A poker-faced gentleman, dressed in black, stood waiting for us holding a sign that read simply, "Ed Colby." When I shook his hand, he picked up my gal Marilyn's bag, and quickly led us to a waiting car.

My memory of the ride into town is a juxtaposition of golden-domed Orthodox churches and severe Soviet-style apartment complexes; road signs in inscrutable Cyrillic; dramatic Communist-era statues; the impatient but orderly flow of traffic; our guide and driver murmuring in Russian in the front seat; the broad Dnieper River; handsome, angular, big-boned Ukrainians hurrying in the streets, stunning women dressed to kill. I said, "Marilyn, we're not in Kansas anymore!"

At a four-story building practically right next to ancient St. Sophia Cathedral, the manager handed us the key to our top-floor apartment. Welcome to Ukraine! Welcome to Apimondia!

Let me make myself perfectly clear: Marilyn and I are not important people. Drivers do not ordinarily pick us up at airports. We just got lucky.

In the summer, as part of my job as a "ranger" on Aspen Mountain, I give weekly lectures on honeybees and beekeeping. At the first one, three Julys ago, I met the most charming family: beekeeper and former Ukrainian President Victor Yushchenko, his wife Kateryna, and daughters Khrystyna and Sophia. Nobody else showed up for my talk, so Mr. President and I did the secret beekeeper handshake, then sat down in the ski patrol hut and talked about our little darlings.

An hour later, I had a personal invitation to Apimondia, 2013, the international bee conference in Kiev. When I muttered something about the cost of such a trip, the first lady said, "I'll find you a place to sleep."

She was true to her word, and then some. We found Champagne and roses on the dining table. And chocolates.

A few days later at their dacha outside of Kiev, Mr. President served his signature fish soup. We dined outside by the lake, all bundled up on a damp fall evening. Second course: pork kabob. I sat next to President Yushchenko. I watched him cut his meat into tiny slices and feed it to the cat under the table.

We were questioned "as average Americans" about our views on the Syrian chemical weapons crisis, then very much in the news. So for a brief moment in time, we represented the whole

United States of America! Marilyn and I were both glad we had an opinion to share.

One of Mr. President's guests informed me that "Ed" was the shortest name he'd ever heard! Ukrainians have impossibly long names, like the president himself, and spellings can vary. So Mr. President's friends call him Victor Andreyovych. Or Viktor Andreyevich. Or simply Mr. President.

Passionate about honeybees, Mr. President keeps 300 hives. I got the biggest laugh when I explained that Michele Obama's single White House hive was strapped down tight to protect it from presidential helicopter prop wash!

Beekeeping is the Ukrainian national pastime. Victor Andreyovych explained that his country has four-and-a-half million beekeepers in a population of only 45 million!

He informed us that the monks at Ukraine's monasteries all make mead. I said, "Well, what do you expect in a country with so many monasteries and so much honey?"

He said, "When you come back next year, you'll have to sample my honey beer!"

After supper Mr. President led us on a tour of his private beekeeping museum – a vast collection that chronicles the evolution of Ukrainian beekeeping – from hollow logs to Ukrainian-style removable frame hives. You never in your life saw so much bee stuff! Long ago Ukrainian beekeepers believed that hives had to be high up in the trees. We looked at a giant wheel used to winch hives into the treetops.

Victor Andreyovych said, "I've taken many heads of state through my bee museum . . . Muammar Gaddafi, Vladimir Putin, Bill Clinton, . . . He mentioned a world leader whose name you'd surely recognize. "How did he like his museum tour?" I wondered out loud.

 "Oh, him? He doesn't like anything," President Yushchenko laughed. Gentle reader, you'd chuckle if I told you who we were talking about. But I'm dancing on thin ice here.

Maybe I was on thin ice the other day when I e-mailed former President Yushchenko some political advice. Was that presumptuous? Because what do I know? But right now – mid-December, 2013 -- hundreds of thousands of Ukrainians confront the police to protest a different Ukrainian president – one who threatens to pull his nation deep into the maw of the Russian bear.

Politics are one thing, the beekeeper bond quite another. Marilyn and I reflect in near disbelief. What extraordinary kindness and hospitality! Two years after an impulsive offer from strangers, I wrote, in so many words, "We're coming. Did you really mean it?" And the reply: "We can't wait to see you!"

In these troubled times, we pray for Ukraine. And count our blessings.

February 2014

Fish soup by the pond with President and Mrs. Yushchenko. Photograph courtesy of Marilyn Gleason.

A gift from Heaven

My gal Marilyn's stuff keeps piling up at the house, and she never really goes home, so I guess you could say she's moved in. I never saw it coming.

It isn't perfect. It never is. Neatness is not her strong suit. She flies around the kitchen trying to use every pot and pan. I clean up in her wake. On winter nights when the alarm goes off hours before dawn, I serve her latte in her nightie. Then she helps me catch the early bus. I arrive home in time for supper and bed. When I'm grumpy, she laughs at me. She takes care of the chickens, and I tend the geese. We never have any spare time.

My race is nearly done. You can't argue with the math. At least I have my health, for now. I can still work. I'd better. My savings largely vanished in the divorce six years ago. I still owe on the farm. None of this matters to Marilyn. I wonder why.

I used to sleep in the ski patrol locker room in Aspen, because my 55-mile commute's so horrendous. My bunk was above the boot dryers. Very fragrant! I called the locker room "my cozy Aspen bungalow," but it never really was. Now I hardly ever stay there. When my co-workers ask why I'd commute so far, I tell them "To get back to those lovin' arms!" They get it. Like dandelions in April, Marilyn's a gift from Heaven.

Of course she loves the bees. What's not to love? It's just that they don't always love her back.

At first, her response to stings was localized swelling. She seemed to be acquiring some immunity. Then, at a bee yard in Mexico a year ago, she got stung twice and suffered a systemic reaction with all-over itching and hives that scared us both. I have to tell you that Marilyn's Irish. She's prone to rashes, blushes, freckles and all the afflictions of those fair-haired, pale-skinned Celts. She's allergic to just about everything.

Marilyn didn't want to give up on the bees, so we decided to take action. She didn't have health insurance, so a series of treatments from an allergist seemed out of the question. But I have lots of bees. We agreed we'd start with a short dose of bee venom.

The first time, she was in the kitchen when I came inside with a bee held between my index finger and thumb. "Hold out your arm," I said. "You're really going to do this?" she gasped. Her eyes went wild as the reality sank in.

"Sure," I said. "Or you could stay home and live in fear of getting stung or even handling my bee clothes and having a reaction. I have the EpiPen if this goes south on us."

"I don't think it will," she said. "You're right. I have to do this."

I gently rubbed the bee against Marilyn's upper arm. At first she was uncooperative. (The bee, not Marilyn!) Then I squeezed her just a little, and she vented her fury. Marilyn gave a little cry, and before you could say, "Jack Robinson!" I scratched out the stinger.

"There," I said. "That wasn't so bad, was it?"

Marilyn was less sanguine. "I'd forgotten how much it hurts!"

She swelled a little. The generalized itching and hives never came back.

There followed many more bee venom administrations. We left the barb in a little longer. Marilyn didn't mind so much now. She watched bemusedly as the stinger twitched and pumped its poison into her hand.

We moved on to multiple stings. I suppose she got stung 50 times over several months. We didn't keep good records. We did this when it was convenient, and each time I gave Marilyn as many stings as she requested.

I don't recommend that you try this on anyone who suffers generalized allergic reactions from bee venom. I never killed Marilyn, although I guess I could have. We talked about it in advance, and she made a decision based on her health care budget and perhaps her commitment to living her life with a beekeeper. We keep an EpiPen on hand in case her throat swells. We live pretty close to the hospital.

And now? I still surprise Marilyn occasionally with a bee in hand. In a weird way she seems to like it. She laughs and comes up with new places where a sting might challenge her immunity. She's apparently as immune as any beekeeper. We'll see.

She's a huge help when I'm moving bees. Now I don't worry about bringing her along. I like having her along. Why wouldn't I? She's a gift from Heaven.

March 2013

Lucky us

Whenever I completely fill a grocery shopping cart with bee sugar, somebody inevitably asks me, "What are you going to do with all that sugar!?"

To which I reply, "I'm going to try my hand at baking some cookies!"

Thankfully the bee feeding is over. Now there's nothing but to wait, and wonder how many hives will succumb to our always unpredictable Colorado winter. Here, a January thaw is a consummation devoutly to be wished, but last year we didn't get one. Without an opportunity for cleansing flights, the bees took sick, and I took losses.

It's always something in this business.

I could have piggybacked my bees onto one of Paul's semi loads to California, and I probably should have, but I hate to let the little darlings out of my sight.

I also don't like the idea of the almond production monoculture. You've seen the photos – blooming almonds as far as the eye can see, and nothing else. Lots of people know about almonds and bees, and if someone asks me if I ship my bees to the Golden State, I'll say, "Almond pollen for bees is maybe like spinach for you. You know spinach is good for you. But if it's all you ever ate, you might take on a greenish pallor. You might not feel so good."

Plus all those bees situated so closely together creates a perfect opportunity to spread diseases and parasites. It's like sleeping around. It might seem like a good idea at the time, but you know that sooner or later you're going to come down with something.

Of course my bees get parasites and diseases anyway, so probably all I'm really doing is making a bad business decision by keeping my bees home all winter. At $150 a hive – or whatever the going rate is – I'm leaving some money on the table. Maybe I'm crazy, but I still hate to let the little darlings out of my sight.

I do want to be smart about money. At 63, I finally have a retirement plan: work until I drop. Before the recession, and the divorce, I assumed that Social Security and savings might provide a modest monthly stipend through my Golden Years. Now all bets are off.

It impresses me that so many people act nonplussed when I say this. No retirement?! These days, people get it. We boomers are not all going to drive our motor homes off into the sunset. But I'm not so bad off. At least I have a job.

And I do enjoy my work on the Aspen Mountain ski patrol. Don't get me wrong: I'm not the patrol director, or even a supervisor. I don't have the talent, or the inclination. I'm down in the ranks.

Patrolling on Aspen Mountain is really a young person's job, so it's either keeping me young or killing me not-so-softly. I suspect the latter. This is a steep little hill, and I sometimes wish there were an easy trail to the bottom.

I signed on a lifetime ago for the thrills – not the money -- but they do pay me every two weeks, and the checks never bounce. The bees help with the bills, too. Every day at work somebody hands me money for honey. My workplace personal interactions complement my honey sales. Honey flies out of my hands.

I guess my dad steered me into both ski patrolling and beekeeping. He worked for the State Department in Washington, later in the Foreign Service, and following that he taught college Spanish. When he worked for the government, he abhorred the bureaucracy and his diplomatic social obligations. At the university, he never fit in politically. He never found a job he loved to wake up to. When I was a boy, he counseled me: "Money doesn't matter. Follow your heart. That's the only way you'll be happy in life."

I took that advice, and I never made my fortune, not even a small one. But I take satisfaction in meaningful work. The bees are part of that.

We beekeepers are so fortunate, mites and Colony Collapse Disorder notwithstanding. We get to pursue an ancient and honorable craft. Who else can say that?

A respected Big Operator spoke at the Colorado Beekeepers' meeting a few summers ago. He said he wondered sometimes why he did it, when life might be so much more satisfying on a 40-hour work schedule.

I thought, "He doesn't get it! He doesn't understand time clocks and work rules and promotions and backstabbing and worrying about whether your tie's on straight." I wondered if he'd ever worked for a company with a human resources department, or received an employee evaluation.

If your back doesn't hurt, if your rig's not broke down, if nobody's yelling at you, if your bees are on a honey flow, if your partner's word is gold, if you're driving down the road in your bee truck, free as a bird, then who should you envy? You're living the dream, and maybe your bees help to make the world a better place.

Our time on this good Earth is brief. Be at peace with that. But we beekeepers made the right choice. We understand what truly matters. We took the road less traveled. Lucky us.

January 2011

The coldest day

Last month you might have read my column about administering stings to my gal Marilyn, to keep up her immunity. I had to submit that piece by January 10th, right in the middle of a cold snap here in western Colorado.

Nine hundred words fills the back page perfectly -- or 800, with a photo. I liked the idea of a bee sting picture, so I opted for the latter. Marilyn was game.

The idea was to show my gnarly ol' fingers holding a bee stinging Marilyn's arm. That's two people tied up. Who's going to take the picture?

After Vietnam, Kelly made his fortune in the sports photography business. I keep bees on his ranch. All I needed to do was get the three of us together, indoors, during daylight hours, in January, with some live honeybees.

Kelly's more than an apiary landowner. He's an old friend, and now he's in love again. He can't stop talking about his new sweetheart.

She has her own ranch, up the valley right next to his. Two ranches, two valleys, two lovers! They're just over the hill from each other, but it might be 10 or 15 miles to drive down and around. They reportedly visit all the time.

Toni's passion is chickens, and Kelly's passion is Toni, so he has a girlfriend, and an egg route. I don't remember how many chickens she owns, but Kelly runs eggs 60 miles all the way to Aspen.

I owed Kelly a case of quarts for my bee yard rent. He kept telling me he'd pick it up at the farm when he made his egg deliveries, but then he never did. So when I called the last time, I said, "I'm going to drop off that honey at your place, and by the way, I need a photo of a bee stinging Marilyn's arm, like today. I'm on deadline."

This piqued his interest, and he understands deadlines. He told me to get up to his place pronto, because as soon as he finished his chores at Toni's, he had to do his own, and then he had to deliver eggs.

So on the morning of January 10 I cracked a beehive and knocked about 30 bees off the underside of the inner cover and into a wide mouth mason jar. The thermometer read five degrees. I didn't really want to do this on the coldest day of the year, but I was pretty quick. And I had a deadline.

When Marilyn and I pulled into Kelly's driveway 45 minutes later, we were 1500 feet higher in elevation. It was just as cold, or colder, than it was at home.

Marilyn taking her medicine. © John Kelly

The bees got all excited from the warmth of the car, so I set them on the walk outside Kelly's house to cool them down. Kelly greeted us at the door, and he put on the Earl Grey tea, which I normally can't stand. But this was *Twinings* Earl Grey, which, Kelly took pains to explain, is superior to all others. This was the original 1720 Earl Grey blend. He said it goes down better with my honey, which is true. It has just the right bergamot, the spice that gives any Earl Grey its distinctive flavor.

All this tea talk, plus looking at photographs of Toni, took some time. I didn't mind. She's easy on the eyes. The bees were still outside.

They cooled down all right. When I checked, they appeared to all be dead, but I wasn't fooled. I brought them inside. In five minutes they were buzzing merrily, so I put them outside to cool down again.

This time I stayed outside and waited until they'd barely succumbed to hypothermia. Then I opened the jar. Cold bees are easy to catch, but they won't sting. I rubbed them repeatedly against Marilyn's arm, to no avail.

This is so simple in the summertime. I just open a hive, pick up a bee, and back her up to Marilyn's bare skin. Marilyn loves the attention.

But this was not summertime. I brought the bees inside yet again, and when they were just beginning to stir, I took them outside and opened the jar. Whenever I tried to pick one up, she flew away! I only had 30 bees! They were flying faster than I could reach for them. On the coldest day of the year, poor little darlings!

Of course I finally caught one that agreed to sting, before they all got away. Marilyn gave a little cry, as the bee tore herself apart disengaging herself from her stinger. All the while, Kelly snapped photos.

We remembered to give Kelly his honey, and we knew when to leave. He had eggs to deliver. I had a deadline.

April 2013

The author's chopped-off 1983 Ford Econoline van, converted to 4WD bee-hauler flatbed.

Photograph courtesy of Ed Colby

A redheaded cousin

Before I got my Rhode Island Reds, I talked to my neighbors, because they've always had chickens.

I wanted a rooster and the rest pullets. I definitely didn't want a bunch of roosters.

He said, "A chicken sexer can tell if the chick is male or female. They pick up the chick and blow up its backside, and that's how they know. Chicken sexers are in very high demand. Not everybody can do it."

"Or would want to," I said.

She said, "I use the nail test, to be sure. You hold the chick in your hand and wait until it's very still. Then you dangle a nail on a string over its head. If the nail goes back and forth, it's a rooster. If it goes in a circle, it's a hen."

I said, "That's the nuttiest story I've ever heard! "

She said, "Look, Ed, you're Catholic, right?"

"Right," I said.

"Then you believe some pretty unlikely things . . ."

"Maybe," I said, "but I'm still don't believe you can sex chickens with a nail on a string."

When I told this story to Father Bob, he didn't seem that skeptical. He said, "You know, my dad could witch a well. He'd hold a forked stick in front of him, and start walking. All of a sudden, that stick would swing down toward the ground, and that's where the water was. How do you explain that?"

I can't explain why I converted to Catholicism and my sister Patty wound up an Evangelical. One of us is Republican, the other not. One lives in the country. The other prefers the suburbs, close to a Wal-Mart. One keeps bees. The other loathes them. Other than these few minor differences, we have a lot in common. We're flesh and blood.

On a perfect May morning, I pick Patty up at the Salt Lake City airport, and we drive to Whitehall, Montana for Aunt Gert's 90th.

We take back roads – through tidy little orchards north of Salt Lake, past ice-blue Bear Lake on the Idaho border, then in and out of Wyoming, skirting west of the Tetons, now up into the timber along the Henrys Fork. I'd like to get out and fish. Pretty soon we're in Montana, close to West Yellowstone, but we take the cutoff to the Madison Valley. I point out to Patty where Dad saw the UFO.

We see our first bee hives, 40 splits set out neatly in the sagebrush south of Ennis. "Oh, just look at the little darlings!" I say.

"Little darlings? That's not what I call 'em!" Patty shoots back.

She had an unfortunate run-in with yellow jackets as a child. I'm sorry.

I was doing a series of experimental bee stings on my bum knee before I left, and I actually considered bringing along some clipped worker bees to continue my treatments, but that would have really taken the fun out of the trip for Little Sis.

We drive over the hill to Virginia City, then down the Ruby Valley to Sheridan. We stop in front of Mom and Dad's old house next to the Episcopal Church on Main Street. They retired here. We drive down to Dad's spot on the Ruby, where he fished almost every afternoon.

Pretty soon we're in the Jefferson Valley. We pass the Barkell Hot Springs bathhouse and pool which is now somebody's home, then the curve where at 19 lovely Aunt Mamie hit her brakes on the ice. She got pinned under the car. Grandma tried but couldn't save her.

Patty spots a cow moose crashing through a hayfield. Hey! A moose? You never used to see moose around here.

We spy Whitehall's red-roofed water tower. On the left is the site of the old Packard Ranch, where Grandma and Grandpa raised seven kids through the Great Depression. They made ends meet with cows, apples, and yes, honeybees.

At the motel, Patty heads straight for bed. Over the phone I hear the wine glasses still clinking at Gertie's, even though the birthday girl has retired for the evening. I can't resist.

Uncle Happy and Gertie raised seven children in a shoebox. Gertie still lives in it. Hap's in a better place. He ran Happy's Bar at the Borden Hotel -- when he wasn't hunting or fishing. When I mention seeing the moose, Cousin Susie says, "There's lots of wildlife now that Happy's gone!"

On Sunday at Top and Gail's, the place is thick with relatives from all over Montana and as far away as Toronto and even Holland. Gail serves Gertie's signature Italian spaghetti. Gert looks all in, but when I ask her if the spaghetti's up to her expectations, she brightens and says, "You'd better believe it!"

My redheaded second cousin Lori wants to know all about honeybees. She's expecting her first bees -- two packages – in the morning. A beekeeping relative! A kindred spirit!

She keeps apologizing for asking so many questions, but I like her spark. Every novice needs a mentor. She can call anytime.

Patty likes to be early for everything, so we take I-15 back to Salt Lake. As we pull into the airport, she says, "This was a good get-together. Everybody got along. Nobody really got going on politics – or religion."

Amen I say.

July 2008

Copper Canyon

The village of Batopilas lies at the bottom of northern Mexico's 6,000-foot-deep, cactus-studded Copper Canyon, where the Mexican timber wolf still howls. Here the principal agricultural business is the business of marijuana. They call it "la mota."

It's late October, but in sub-tropical Batopilas at 1500 feet, it's hot, and the streets are a riot of strange flowers.

I don't see honeybees on the flowers. Butterflies appear to be the main pollinators – orange ones and yellows and grays with long tails.

Bees do hover around food stalls looking for a taste of cut fruit or spilled soda pop. They act like yellow jackets. You never saw anything like it.

To walk into the Casa Monse hotel is to walk into the Garden of Eden. Hammocks swing from lemon trees in the courtyard. Thirteen-year-old Julieta roasts tortillas on an open fire, while a honeybee hovers nearby. Pungent tropical smells -- sweet strong essence of damp earth, and of life, and death -- envelop you.

The Casa Monse is not a first-class hotel. When I walk in for the first time, I have a touch of fever. Senora Monse greets me like the Prodigal Son. She is even older than I am. "You are Eduardo," she says.

There's something about this woman that I like. Luckily, I speak Mexican. "How would you know that?" I tease. "You must be a witch."

Her dark eyes twinkle as she hooks her arm 'round my waist and leads me to my room. "The Spanish women told me you were coming," she says.

I ask Senora Monse about local beekeepers. "We had one," she says. "He used to come around, but he died."

Faintly delirious, I sleep spread-eagled under a big ceiling fan all through the afternoon.

That evening at the cantina I join Carla and Eva, charming veterinarians from Barcelona. Their lisping, lilting, murmuring Old World Spanish confounds me. They say the Mexicans talk funny. It's a rowdy crowd this Saturday night. A Chihuahua cowpoke sings lonesome cowboy tunes on the guitar. Carla croons the Beatles' "Yesterday" in Catalan. Eva says, "We saw a bee today."

As we linger over Indio beers, the girls confide that the Casa Monse is plagued with scorpions. "You need to pull your bed away from the wall," Eva says. "And look under the covers."

Next door to the Casa Monse, at the newer Hotel Juanita, the proprietress – one Senora Juanita -- smiles when I ask about honeybees. She points to her kitchen windowsill littered with dead bees.

"I kill them when they come into my kitchen looking for food," she says. "We have lots of bees."

The actual keeping of bees is apparently dead here. At the old stone hacienda ruins across the river, pigs root in the halls. I knock on the door of a house built on the hacienda grounds. I am searching for a man reported to be a former beekeeper. I am looking for the uncle of Senora Juanita, for Senor Manuel Acaraz.

I hand Senor Acaraz my beekeeper card. "Ahhhh," he says, "you are the beekeeper who came looking for me yesterday."

Nearly toothless, and advanced in years, Senor Acaraz's eyes light up when he talks about bees. He kept a dozen hives, until the Africans invaded these canyons.

"My bees began stinging people across the river," he says. "They harassed the tourists visiting the hacienda. The African comes out fierce, and she won't make honey. Who would want these bees?"

We lean against his pickup truck. Sr. Acaraz is clearly enjoying the conversation, as am I.

"How about European bees?" I ask. "Are there any left?"

"There must be," he says, "because from time to time people bring wild honey from high up in the canyons. They say it comes from gentle bees. Sometimes the colonies are very large."

I wonder if the higher elevations of the canyon – up to 8,000 feet – where it snows and freezes hard, might offer a refuge for non-Africanized bees – a place where their cold-hardiness offers them at least one advantage over the Africans.

That's pure speculation.

At an orchard at 5,500 feet, the farmer and I talk apples, until the conversation inevitably turns to honeybees.

"The African is worthless," he says. "In the spring I caught a swarm and put it in an apple box, but most of the bees left. Now there's just a little bit of comb on the outside of the box."

"Did you wear a veil to catch the swarm?" I ask.

He smiles. "I don't have one," he says.

I ask if I might take a look.

He'd thrown some plastic over the apple box, "to keep them warm," he says.

He pulls up the plastic and says, "Stick your head under there."

I don't have my veil, but naturally I do what he tells me. Looking up, I see a pancake-sized piece of bee-covered honeycomb 24 inches in front of my face. One of the little darlings bounces off my eye socket.

"These are all that are left," he says. "How are they going to make any honey?"

I shake my head. He says, "Do you think there's a queen in there?"

"Maybe," I say.

Back in Batopilas, Senora Monse asks me if I will be staying another night.

"No," I say.

"Well, where are you going?" she asks.

"To the Casa Juanita," I say.

Her eyes shoot daggers. "What happened last night?" she demands.

"It's the scorpions," I say.

She says, "You saw a scorpion?"

"More than one," I say.

Senora Monse looks to her frail and ancient husband. "Our guest reports that he saw a scorpion," she says.

The old man acts genuinely surprised. "A scorpion? We never had one before," he says.

January 2007

A-fib

Here in Colorado, the Division of Wildlife provides beekeepers with solar-powered electric fencing to keep the bad bears out of our apiaries. These fences pack a wallop. I know.

Until recently, it never occurred to me that a self-induced fence shock might save me a trip to the hospital.

My cozy Aspen bungalow doubles as the Aspen Mountain ski patrol locker room. One early morning last winter, before the crew arrived, I savored a cup of joe. Suddenly felt I flushed, weak, uneasy. I thought this was a reaction to caffeine on an empty stomach.

I went up the lift with the rest of the ski patrol, but I wasn't right. Finally I said to one of the patrol paramedics, "Hey, Alex, you want to feel a strange pulse?"

After palpating my wrist, he said, in the most casual way, "Mind if I hook you up to the (heart) monitor?" Paramedics live for moments like this.

After a few minutes, he calmly announced that I was in "atrial fibrillation," or "A-fib," which means my heart was beating erratically, and way too fast.

I couldn't comprehend this, because I am in perfect health. Always have been.

One of the ambulance paramedics peddles my honey to the hospital staff. He does it to meet chicks. At the emergency room, once word got out that I was the "bee man," doctors and nurses and lab techs came out of the woodwork to introduce themselves and tell me how much they liked my honey. I felt like a celebrity.

The ER staff gave me every cardiac test known to man. The cardiologist told me I could have a stroke if I didn't convert back to a regular rhythm, because blood could pool and clot inside my fluttering heart.

He also said, "How much did you have to drink last night?"

I said, "Three or four beers, max."

He nodded sagely. "Ah, the 'holiday heart' syndrome," he said. "Lots of people experience their first episode of A-fib on the heels of a hangover."

The doctor didn't seem to get it. "I didn't have a hangover," I said. "Coffee set this off."

He shook his head. "It wasn't the coffee. It was the alcohol," he said. "It has a delayed effect."

There's no point arguing with doctors.

They pumped me full of medications, and the doc said, "You'll probably convert (back to a regular pulse) sometime tonight. Come back in the morning. If you're still in A-fib, we'll shock you."

Shock me? Why? I'm in perfect health, and I'm only 60.

That night at a Mexican joint in Aspen, I kept checking my pulse. The medications had brought my pulse down to a normal rate, but it was still irregular. My heart would seem to beat normally, and I'd think I'd converted. Then it would skip a beat, or I'd suddenly get two or three right beats together. I took my pulse so many times, I forgot what a normal beat felt like.

My buddy Doug said, "Here, let me try."

I extended my hand across the restaurant table, and Doug earnestly held my wrist in his hand. Suddenly we heard a familiar voice from a few tables away. "Hey, what's going on over there?" our friend Karen wanted to know.

"Doug just proposed to me!" I exclaimed. Sometimes you have to play along.

After supper, I tried some tips from my co-workers. I sent my pulse skyrocketing as I walked briskly up the hill to my bungalow. I took an ice-cold shower. I strained like I might on the toilet. I massaged my carotid artery. All to no avail.

My condition remained unchanged through the night. In the morning I reported back to the emergency room. By now I had another concern – my medical bill. The young man at the desk wanted to re-admit me to the ER. My insurance has a $500 co-pay for each emergency room visit. I said, "Whoa! I was just here yesterday, remember? I feel like I never really left. I only need a little shock to my heart, not a full workup."

He said, "That's not the way we do it here."

Fortunately I knew the head nurse, who can be belligerent. The cardiologist was also on my side. It still took the two of them a couple of minutes to talk sense into the young man behind the ER desk.

Next I said to the cardiologist, "How about if we skip the anesthesia?"

He looked amused. "Why would we do that?" he asked.

"I'm trying to keep from blowing my life savings," I said. "I don't need anesthesia. This only lasts a second, right? How much can it hurt?"

"A lot," he said. "Plus, they'd pull my license."

After the charming cardiology nurse hooked me up to an IV dripping a creamy-white knockout drug, she cooed to me as she gently stroked my arm. That's the last thing I remember.

When I awoke, she was still there. I looked at the cardiac monitor. My heartbeat was regular again. I said, "Scarlett, did I jump like a shocked CPR patient?"

"Oh, yeah!" she said brightly.

Looking back, I suppose it was all worth it. I didn't have a stroke. I supposedly have my A-fib under control with medication now. I can whoop it up with Spartan moderation. I hardly remember what coffee tastes like.

But I really didn't want to get shocked at the hospital. I did everything I could think of to convert on my own, but I never thought to shock myself with my electric bee yard fence! I can't understand why it never occurred to me.

I'd have tried that in a heartbeat.

December 2007

The longest night

When a cool evening wind blew the bees home early, I knew they'd stay snug in their hives all night. I mused that I might lay my pillow against one of the landing boards, and that the little darlings' humming would lull me to dreamland under the stars. But it looked like it might rain, so I pitched my tent instead.

It was the end of May, and I was on the first leg of a two-and-a-half-day beekeeping odyssey. The sun was already low when I arrived at the Elk River Valley, up by Steamboat Springs, Colorado. There'd been a problem last year with a bear in this yard. I put up a woven wire fence and connected it to an electric charger and solar panel. It ought to stop a bear, but you never know. I figured I'd work the bees in the morning.

I'm leasing bees from Jack this summer. They'd finally gotten moved back to their summer yards after pollinating the almonds in California. This was my first trip to this particular location. The landowners live across the road. I wanted to introduce myself, but when I finished the fence, I thought it might be their suppertime. But then if I waited, it might be their bedtime. Unsure what to do, I retired for the evening.

To be perfectly honest, I wasn't totally comfortable with my decision to sleep with the bees. It occurred to me that if the bear smelled honey and brood, I'd be right at the source. Still, I'd be safer inside the electric fence than outside. But I felt like one of those guys who goes down into the deep in a shark-proof cage after they chum the water with blood.

My $10 sleeping bag that I bought at the grocery store would have been the perfect size for a Cub Scout. The only way I could cover my shoulders was to sleep curled up. When I woke up, it felt like it had been a long night. I thought it must be nearly dawn, but when I looked at my watch, it read 10 p.m. I turned on my flashlight and read an article about pest management strategies for Varroa mites. I almost finished it.

Just outside the tent, the fence charger "snap-snap-snapped" reassuringly all night long. I wondered about African campers who ring their campsite with fire to keep away the wild beasts. I thought about a young local journalist who went to Africa to seek adventure. Out in the bush, he at first thought it odd when his guide slept in the Land Rover. In the middle of the night, he heard lions padding through the camp. I dreamed that my tent touched the electric fence, and that I could feel electricity tingling in my toes. I didn't dream about the bear.

In the morning, ice coated the tent walls. Clouds and wispy strings of mist hung low over the wide valley, and the fields shimmered white with frost.

I got up early. My bee yard campsite was on a gentle sagebrush-covered hillside in full view of the ranch house across the road. When I performed my morning ablutions, I appreciated the privacy afforded by the shelter of my flatbed Ford, and I was glad I'd thought to bring a shovel.

By the time I finished supering the bees, the rancher came by on his four-wheeler. He introduced himself as "Doc," although he said he wasn't one. He peppered me with questions about honeybees and couldn't have been friendlier.

Later, when I used the phone at his house, he said, "You should have stopped by last night. You could have slept in the bunkhouse."

It's a little backwoodsy at the big yard over by Hayden. They raise chickens, sheep, goats, ducks, geese, turkeys, and dogs – lots of dogs. Critters wander around everywhere.

You see a lot of sheep guard dogs in this country. Mainly they're Great Pyrenees, Akbash, or Anatolian Shepherds. They're all big white dogs, and they're mostly friendly when they're not defending the flock.

As I introduced myself to the owner, a wolf-sized dog nuzzled my hand. I said, "What kind of dog is that, anyway?" The man said, "Three quarters Anatolian and one quarter Pyrenees. No wait, I guess he's two thirds Anatolian and one third Pyrenees."

I didn't dispute the math. I said, "What breed do you like the best?"

The man said, "The Anatolian. They're the best bear fighters."

I said, "Excuse me, but I thought you just said 'bear fighters.'"

The man said, "Sure. This guy will take on any bear. Bears don't come around here."

The wife said, "They might come around, but they don't leave – not alive."

This sounded to me like a good place for bees. And camping.

Pointing to a big sad-eyed dog with swollen teats, the woman said, "That Pyrenees bitch killed 15 of my chickens last night. Fed them to her pups."

"Oh," I said.

The man said, "We hauled a road kill deer up by the bees – for the dogs -- but I don't think they'll bother you tonight."

I said, "They won't bother me. I like dogs."

Three off-white pups rolled in the dirt at our feet. The wife said, "I've got two more litters.

Do you need a dog?"

"Got one," I said. "But thanks, anyway."

As I got ready to head up to the bee yard, the woman looked at me hopefully. I guess she figured she had me hooked. She said, "When you leave, you be sure and take one of those pups."

August 2004

Swarm-catching. Photograph courtesy of Marilyn Gleason.

An Amish snowboarder

My gal Marilyn is such a schemer. I told her, "The Medina County beekeepers want to fly me to Ohio to give a talk."

"Fly you?" she shot back. "What am I supposed to do? Stay home and feed the chickens?"

Twenty-four hours later she had a plan. "Look, let's take the train. We'll get an eight-stop, 15-day Amtrak pass. We can get off in Cleveland for your talk. Whatever it would have cost them in airfare, have them just give you the money, and we'll use it for train tickets."

I told you she's a schemer.

In October we boarded in Grand Junction, Colorado. The California Zephyr comes right up the Colorado River valley past our house. You can see it from the train. When I caught a glimpse of my big New Castle bee yard, I momentarily panicked and wondered if I'd turned on the solar electric bear fence.

We got off in Glenwood Springs to stretch our legs, and the conductor warned us to stay on the platform. After a few minutes, Marilyn said, "I'll be right back," and disappeared.

I thought nothing of it and went back to the observation car. When the train left the station, I didn't see Marilyn, but she can take care of herself, right? At first I didn't worry. I had her cell phone. But after a few minutes, I headed back to our coach seats, just to be sure. I asked the woman across the aisle. No Marilyn. I checked downstairs.

When I approached the conductor, he immediately began lecturing me about staying on the platform during "smoke break" stops. Lectured *me*! About then a helpful assistant conductor cut into his harangue and offered to make a PA announcement.

"We have a missing passenger. Marilyn Gleason, please check in with your traveling companion in the observation car," she announced. The train instantly pulsed with excitement. A passenger left behind?! I tried to stay calm as I wondered how I'd kill a day in Denver waiting for Marilyn to catch up. Everybody was looking at me.

A minute later when Marilyn showed up beet-red, the observation car broke into applause. She'd made a new friend in another car, that's all, and you know how it is when you get to talking.

We woke up the next morning in Iowa, where corn is king. I reflected that that all those corn stalks came from seed dipped in neonicitinoid systemic pesticides, and that with corn prices that topped out at $8/bushel, a lot of pollinator habitat got converted to corn.

We got off in Cleveland, which is close to Medina. The Medina beekeepers put us up in the charming but haunted Spitzer House bed and breakfast. The B&B was filled with antiques and dreamy Midwestern landscape paintings that make you yearn to be young again and smoke a corncob pipe and head off down the river with Huck Finn and Jim.

Medina is the Holy Grail of American beekeeping. It was here, about 1865, that jewelry manufacturer A.I. Root paid a man a dollar to catch a passing swarm of honeybees. Root became so intrigued with the little darlings that he founded a beekeeping empire. In short order, this beginner beekeeper progressed from honey producer to manufacturer of centrifugal honey extractors, smokers and then-revolutionary Langstroth woodenware. He initially wrote how-to articles for the American Bee Journal, before creating his encyclopedic ABC of Bee culture (later ABC and XYZ of Bee Culture) and founding his own beekeeping periodical, Gleanings in Bee Culture.

Today the Root Company centers on publishing and candle-making. Bee Culture magazine editor Kim Flottum showed us through Root's candle-making factory, housed in the same building A.I. Root built next to a railroad hub over a hundred years ago. This operation employs robots but also flesh-and-blood Americans. Made in Medina, not China.

Later in the trip, in the Marie Reine Du Monde Cathedral in Montreal, Marilyn picked up a votive candle. "This is a Root candle," she said.

I talked to the Medina beekeepers about sideline beekeeping. May I summarize? Control your mites. Don't put all your eggs in one basket. Don't sweat the petty stuff. Embrace failure, the greatest teacher. Don't quit your day job. We all sang a song about honeybees, and then Marilyn and I slipped away into the night. We talked about Medina for the rest of our trip, because how often do you get treated like royalty?

Next stop: New York City. You never saw so many skinny people! Marilyn showed this country bumpkin around. We walked in Central Park, lunched in Harlem, dined with old friends. Kim gave us the name of a rooftop beekeeper enthusiastic about showing us how they do it in the city, but our trip was too brief, our planning too pathetic. Jim was just leaving town. But he offered to buy us a drink. This beekeeper comradeship made me feel all warm and fuzzy.

The train was thick with Amish. They can be outgoing with strangers, even chatty, when they're not on their cell phones. On the trip back I met a lanky Amish young man who farms 20 acres with horses and whose "favorite thing in life" is snowboarding. He also works for an Amish contractor installing automated feeding systems for factory hog and chicken farms.

He was on his way to hunt elk in Telluride. Beekeeping piqued his interest. He asked how much I make on 100 colonies.

We arrived home in Colorado at the end of an October heat wave. Bees were busy devouring their winter honey stores. Vacation was over. Back in the saddle, again.

January 2015

An unforgettable woman

I generally do pretty well selling honey at the early-November Aspen ski patrol medical refresher. This two-day affair draws a couple of hundred attendees. I pitch product on day one, so folks can bring money the day following.

At the last refresher, I ran into the ski area vice-president, an amiable guy who seems amused to have a beekeeper on the payroll.

He said, "My beekeeper friends on the Eastern Slope had a fabulous honey harvest. They have acres of alfalfa outside their door and no competition from commercial beekeepers."

He's mentioned these people before. They sometimes send him my columns when they touch on the operations of the Aspen Skiing Company. So I have to watch what I say!

I said, "I know a few beekeepers. What are your friends' names?"

The woman's name meant nothing to me, but the man's hit me like a gut punch. It was a name I'd never forget. It was the name of a man I'd never met. Long ago, he married Beth.

In 1967, when I was just a kid, I loaded up my Rambler station wagon, drove over Loveland Pass in a snowstorm, and landed a job washing dishes at a ski lodge in a new resort called "Vail."

That's where I met Beth. She might have been a bus girl. I don't remember. I think she was bowlegged, but maybe that was somebody else. I'm sure she was a tomboy. We skied together. I was hopelessly smitten. This was unlucky for me, because she was engaged to a Marine pilot. The Vietnam War raged. I'd have wound up over there myself, if I hadn't already flunked my physical.

So there wasn't any smooching with Beth. But we hung out. I adored her. I didn't want that winter to ever end. When it did, I hitchhiked to California to visit her at her parents' seaside home. We went to Disneyland.

Sixteen years later I ran into her working in a grocery story in Boulder, Colorado, where I was attending the University of Colorado. She'd married that pilot. They had a family, and she had stage four cancer. She seemed inexplicably upbeat. We thought maybe we'd get together for lunch, but we never did. After awhile, I stopped seeing her at the store. I figured she was a goner. I thought about calling, but it felt awkward. "Hello? Are you Beth's husband? I'm oh, just a friend, and I was wondering if she's still alive . . ."

All these years I wondered, until last November after the medical refresher, when I called the Marine pilot beekeeper whose name I still remembered after all these years.

He was a little surprised to hear from a Bee Culture columnist. I got right to the point: "Is Beth still on this Earth?"

"Oh, sure," he said matter-of-factly. She'd made a "miracle recovery," he explained -- after refusing chemo and radiation. Later, they divorced. They both re-married. Now they both live in the country. They both have bees. She has horses. The kids are grown. He sounded at ease explaining all this to a stranger on the phone.

I took a deep breath. Life never turns out the way we expect. All I wanted now was for all of us to live happily ever after.

Back at the patrol refresher, I made a pitch for improved health through the regular consumption of honey. You have to believe in your product! I explained Dr. Ron Fessenden's argument that the perfect ratio of fructose to glucose in honey acts to promote the production of glycogen in the liver, which keeps your brain fed through your nighttime fast, eliminating one cause of insomnia.

I got some skeptical looks from the audience, but I ran this past a freethinking doctor friend before I gave my little speech. "Sounds reasonable to me," he said. Plus I freely admitted a clear conflict of interest, which got a laugh from the crowd.

I sold plenty of honey, but only one person expressed an interest in getting more sleep by eating a teaspoon of honey before bed. One person out of 200? This is such a no-brainer! This is the best-tasting medicine you'll ever take. Why wouldn't you at least give it a try?

I asked my insomniac honey buyer to let me know if the treatment worked, and so far the answer is no. Of course I never said it would. I only said give it a try. It works for me, but I have an advantage: I believe. Maybe it's all in the mind!

Of course Beth was alive all along, but in my own mind, I brought her back from the dead with a lucky phone call. I'm pleased she beat the odds. It gives me peace knowing she's still on this green Earth. I hope she's happy. I suspect she is. I've got her number. One day I'll call.

March 2014

Boiled lobster, with an itch

In Don Juan's Yucatan jungle bee yard, the little darlings were making a little honey, but they weren't in a great mood. Guard bees peppered our veils as soon as we walked into the apiary.

Don Juan typically eschews gloves when he works his Africanized bees, so we followed his lead. Plus my sidekick Marilyn was in charge of taking pictures, and gloves would have made her job more difficult.

Most of these bees weren't pure African. Don Juan and his partners in his *ejido* co-op continually introduce Italian queens to keep their stock manageable. These girls were a little testy, but I've worked with Carniolans on a bad day that weren't any harder to deal with than Don Juan's African hybrids.

At a colony that seemed particularly annoyed when we walked by, bees were bringing in an interesting white pollen. "Get a picture of that hive entrance, would you?" I said to Marilyn.

I didn't pay too much attention to Marilyn. Don Juan and I were preoccupied with bee talk and brood patterns.

Marilyn is new to bees but not to allergic reactions. She breaks out in hives for a variety of reasons, including exposure to new-mown hay and sneeze weed. She had a reaction pruning apple trees. At least she's not allergic to me.

Typically she starts itching on the insides of her elbows. Then it spreads like wildfire, as she goes into a delirium of itching and scratching.

But with bee stings, the worst she'd experienced were severe localized reactions – like the stung hand that swelled so big it kept her up for two nights running. Yet like a seasick sailor who nonetheless loves the sea, Marilyn was determined. She kept coming to the bee yard, and eventually her bee sting reactions diminished. She thought she had this behind her.

Taking pictures of those bees bringing in white pollen in Don Juan's yard, Marilyn got "*Ow!*" stung at the base of her thumb. She didn't think too much of it. Then "*Ouch!*" she got it in the exact same place on the other hand. Maybe she said a bad word. I would have. She kept on taking pictures.

All of a sudden she didn't feel so great. She left the bee yard and took off her veil.

She felt "weird," she said later. But she put her veil back on and came back out to the apiary.

When our visit with Don Juan's bees was over, Marilyn said, "I got stung. I'm starting to itch everywhere."

It was hot. She was drenched in sweat. Her face was red and blotchy, and she was scratching like some Mexican dogs we met on our trip.

"Your throat isn't swelling?" I asked.

"I'm fine," she said. "But my feet, hands, elbows, wrists, head, and rear end all itch like crazy. I just need some Benadryl."

I wasn't convinced. One danger of a systemic reaction, like hives, is that it can be the precursor to throat swelling that could block off a windpipe and lead, ultimately, to a dead Marilyn. We didn't need that.

Benadryl reduces itching and hives, but to save someone's life when she's struggling for air you need epinephrine.

The nearby town of Xpujil had a pharmacy and a hospital. Don Juan navigated for us as I hot-footed it into town in the rental car. We soon had Marilyn's Benadryl. She took it right there in the pharmacy.

Our little cabana was close by the hospital, so we decided we'd give the patient some rest, and the privacy to do some real scratching. If her throat started to swell, we could head for the emergency room.

If we'd have thought of it, maybe we could have bought an "Epi-pen" at the pharmacy – a measured dose of epinephrine that a patient can self-inject in the event of difficulty breathing during an allergic reaction. Here in the U.S. this requires a doctor's prescription, but Mexico's prescription laws tend to be more lax.

But Marilyn insisted she'd be fine. While this was the most dramatic case of hives she'd ever experienced, she'd had hives before and never had the swollen throat problem. So why would we borrow trouble now?

Meanwhile, her condition went wild. She turned red all over. Stretched out on the bed at the cabana, she looked like a boiled Maine lobster, with an itch.

Then she got better.

The only allergic reaction I ever had was to shellfish. First I had itching and the hives, followed by the swollen throat -- a condition that resolved itself. Now I have an Epi-pen, which I should carry with me at all times but never do. Since my unfortunate episode, I've twice consumed small quantities of shellfish by mistake, with no ill effects.

Marilyn is a very competent mechanic! Photograph courtesy of Ed Colby.

As for Marilyn, I guess it was the severity of her reaction, combined with its being brought on by a bee sting, that most alarmed me. It doesn't seem like you could die from something like Marilyn's allergic reaction brought on by pruning an apple tree, although I suppose you could. A bee sting is different. It's the classic trigger for a true medical emergency.

As an experienced allergy sufferer, Marilyn's take on this is interesting. She feels that the stifling heat of the tropical bee yard might have pushed her over the edge. Maybe that's wishful thinking, because she likes working with bees.

We don't have a plan for the future, yet. Maybe bees aren't in the cards for Marilyn. She doesn't want to quit, and I like having her out there in the bee yard with me, but we don't need any more medical emergencies. Marilyn's a great gal. I'd miss her.

January 2012

Broke down in Steamboat

I go up to Steamboat Springs, Colorado to chase bees. It takes me about three days to make my rounds. I sleep in the bee yards. Then I head back to the honey house in Meeker and back home to the farm in New Castle – a total of about three-and-a-half hours from Steamboat, the way I go.

I love my 1983 Ford one-ton honeybee truck, even if it's ugly, and even if that 460 engine does like to drink gas. Let's not talk about gas mileage. But heading up the hill out of Steamboat toward Clark the other day, the beast up and died. It did this without warning or provocation. I coasted to a ranch driveway and pulled off.

Any combustion engine needs two things – fuel and spark. I disconnected the gas line at the carburetor and turned the key. Then I peered under the hood. Gasoline was boiling on top of the engine.

This didn't strike me as a good time to check for spark. Everything in my ignition system was brand new, except for the coil (the electromagnet that intensifies the spark that burns the gasoline) and the control module -- the minicomputer that regulates ignition. I didn't think about the control module right then. I haven't owned this truck very long, and I'm not much of a mechanic. To be perfectly honest, I didn't know I had a control module.

It was 4:45 on Saturday, and I decided to call an auto parts store pronto and buy a coil. I had to do something, even if it was wrong. When I walked up to the ranch barn to use the phone, the from-out-of-state ranch owner said, "It's probably your timing chain. They'll have to tear your engine apart." He seemed to take pleasure in telling me this.

I said, "I don't want to hear that. I need to fix this right now."

The lady at the auto parts store said she'd leave the coil in a plastic bag hanging on the bumper of the trailer outside, because they were closing. Afterward the ranch owner said, "It's not your coil. It might be your control module. But it's probably your timing chain."

"Well, can you recommend a good mechanic in Steamboat?" I asked.

"I can't," he said, "because there aren't any. You'd have to go to Craig."

"How would I get my truck to Craig?" I asked.

"You can't get anything done in this town," he said.

At this point I called Esther and threw myself at her mercy. "Esther, rescue me," I said.

"I'll be there in ten minutes," she said.

Esther and I go way back, although I never really knew her that well. She's the daughter of my very old and very dear friend Granny, who in my youth taught me to smoke and cuss. Granny kept telling me I ought to look up Esther in Steamboat this summer, and I'd really meant to, but you know how it is.

Now I felt a little awkward calling her simply because I needed help. I wished I'd made a social call first – taken her to lunch, whatever.

Esther put me in her grandkids' bedroom downstairs. I slept in one of those narrow little kids' beds. The room was full of toys. "You can sleep with a teddy if you want," Esther said. The next morning over coffee and eggs, I got her laughing, which is something I can sometimes make people do.

She loaned me her car. I installed the new coil, but the truck still wouldn't start. I managed to shock myself twice testing for spark, so I knew I had it. Now the only thing between the repair shop and me was a control module. After I bolted on a new one, I was almost afraid to turn the key, because this was my last glimmer of hope, and I had a sinking feeling.

Occasionally things go my way, however. The engine started right up.

Earlier, when I didn't really think I'd get on the road that day, I'd made plans to lunch with my old ski patrol buddy Wilbur. Now, with the truck fixed, and after nearly a week on the road, I was itching to finish up with my bees and head for home. I can be self-centered, and in fact that is my nature. But this one time I put a friend first.

You're probably wondering if I had honey supers on the truck, and the answer is "yes." Fortunately, there was a honey flow, so robbing bees didn't plague the supers while I was parked in a stranger's driveway. This could have complicated things, especially if I'd needed to get towed to a garage.

So my tale has a silver lining: No robbers, and the whole adventure only cost me a hundred bucks – well, maybe a couple of hundred -- and I got to renew two old friendships. How big a disaster is that?

Plus now I've replaced everything in my truck's ignition system. If it quits on me again, I'll know it has to be the timing chain.

A week to the day later, on Saturday evening, I lost a "dually" wheel hauling three tons of honey through Craig. This time another friend bailed me out, and the weekend cost me a lot more than $200 . . . but it's another story.

October 2004

Bad luck in Craig

I had some trouble this summer with my old one-ton flatbed Ford. My bad luck peaked one Saturday evening when I lost a "dually"wheel while hauling a full load of honey supers through Craig, Colorado.

I won't say it was all bad luck with that truck. It had a few shining moments, like when I rebuilt and re-jetted the carburetor, and the gas mileage jumped from six mpg to 10.

But I was lucky I didn't kill anybody when that wheel came off.

I'd just looked at the wheels, or maybe I should say the tires. I was concerned about tire pressure and overloading the truck, because those were pretty full honey supers. At a highway rest stop, I confirmed that the tires looked fine. I just never thought to check the lug nuts, which must have been ready to come off.

I drove 55 all the way from Steamboat Springs to Craig – totally oblivious to my peril. But maybe I have a guardian angel, because it was on the 35 mph city bypass that I suddenly realized something was horribly wrong.

OK, the wheel didn't literally come off the truck, but it sheared off the axle studs, so the wheel wasn't attached to the axle. Had I been cruising down the highway, the wheel might have gone flying off. I suppose it would have.

A Good Samaritan followed me to the side of the road. He said, "Oh, man, you're not going anywhere now, are you?"

When the Samaritan offered me his cell phone, I called my partner Jack in Meeker. I said, "Jack, I've got some good news, and I've got some bad news. Which do you want first?"

Jack said, "I'll take the good news."

I said, "I'm still alive."

"Obviously," Jack said. "What's the bad news?"

I said, "I need you to come down with your trailer so we can get this honey off."

Jack's a morning guy. He's not much use after about sundown. He started muttering, but he calmed down when I said, "Jack, you don't have to come right now. Come in the morning."

"In the morning? Oh, all right, I'll meet you at 8:00," he said.

I noticed the Good Samaritan swatting at something, and I wondered if it was mosquitoes. He flailed his arms and bobbed his head like he was getting dive-bombed by something.

Then it hit me. Until that moment I frankly hadn't noticed any bees flying out of the supers.

I said, "You don't want to swing at those bees. They won't bother you."

But I was wrong, because they already were bothering him. Now I have to give this guy credit. He really felt he was in danger, and yet he stood by me – a stranger in need. That's my definition of heroism.

He drove me to the Trav-O-Tel, on Victory Way. He said, "Does this place look OK?"

There were geraniums in the flower boxes, and a new metal roof extended over the walkway in front of the rooms. A group of Mexicans sat outside with their shirts off drinking beer. I said, "This is perfect." When the nice desk lady told me a room was $34, I was so pleased I never even inquired about my AARP discount.

The room was small but clean. It had pink wallpaper with bunnies in wheelbarrows. The TV worked, and there was plenty of hot water.

Just down Victory Way, on the sign over the front door of the Popular Bar, a well-endowed blonde leans over a pool table to make a bank shot. Inside, the man at the end of the bar said, "If Bush gets re-elected, this country will be nothing but the rich and the poor. There won't be a middle class." Then he fell off his bar stool.

At breakfast the next morning at the Golden Cavvy Restaurant, I asked the waitress what a "cavvy" was. She gave me a look that said if she got asked that question one more time, she'd kill. "It's a herd of horses," she said. Didn't I know anything?

Signs on the wall read, "No liquor before 7 a.m." and "No hamburgers served under medium well – The Management." This was my kind of place.

Jack and I got the honey onto his trailer. A friendly cop with "Rebel" tattooed on his arm called his buddy who did emergency road service. Soon I was mobile again, but just for temporary.

Both left rear wheels needed to be replaced, along with the hub. It was Sunday morning, so I resigned myself to another night at the Trav-O-Tel.

Back at the room, I met my neighbor in the room next door, an old cowpoke visiting his daughter in town. "She and I don't see eye to eye," he said.

I had the truck parked in front of his door. The cowpoke said, "What's that smell?"

I explained to him about Bee-Go and asked, "Does it bother you?"

He said, "Let me put it this way. I'd move that truck before the manager asks you to."

By the time I paid my repair bill the next morning at the junkyard, I was out over $500 for the weekend. But what are a few expenses when your bees are in a big honey flow? Honey prices were still sky-high. I was on the road again. And I'd learned an important life lesson: always check your lug nuts.

December 2004

Cherry biscuits

"Every day's a great day to be alive!" That's what Dave told me when I dropped off bees to pollinate his sweet cherry orchard last March in Grand Junction. I appreciated his optimism. In the Colorado fruit business, you need optimism, or a psychiatrist. Last year Jack Frost came calling at Dave's place. "I got just enough cherries to make a couple of cherry biscuits," Dave quipped.

Driving back into town from the orchard, I got pulled over by the state patrol. I wasn't even on the highway. This was a country road. The good-natured cop said, "Good morning! I noticed you crossed the center line a couple of times, and I was wondering if maybe your dog was distracting you."

Crossed the centerline? What was he talking about? I was just driving down the road, dreaming about honeybees!

My gal Marilyn's blue heeler dog bouncing around in the cab should have made a perfect setup to tell a harmless white lie, but I'm not a good liar, as Marilyn will attest.

"I guess I don't have a reason," I blurted.

"Well, be careful. I'm going to give you a warning," he said. He glanced back at the hand truck, ramp, smokers and hive net in my truck bed. "What do you got in the back of your truck?"

"Bee stuff," I replied. "I just dropped off some hives at an orchard in the Redlands."

The officer's eyes lit up. "My wife wants bees," he said. "But she doesn't really know anything about them."

"Beekeeping can be a little overwhelming at first," I said. I gave him my beekeeper card. "Tell her to call me anytime," I said.

Now I wonder if he pulled me over just to strike up a bee conversation. Because I'm a good driver. I always stay in my lane.

At breakfast at the Slice o' Life Bakery in Palisade, a dozen women chatted at the next table. When I got up to leave, my plate inexplicably jumped out of my hand and flew across the room. It bounced a couple of times before it wobbled noisily to a stop on the floor, unbroken. For an instant, you could have heard a pin drop. One of the women broke the silence. "Oh, my," she said.

I said, "This morning I got pulled over by the highway patrol and didn't get a ticket. I just threw my plate across the floor, and it didn't break. This is shaping up to be a good day."

I do like the Slice o' Life Bakery. You walk past a native-bee poster on the way in. It reads, "Our Future Flies on the Wings of Pollinators". How could you not love this place!? The diminutive long-time proprietress, Mary, hops around the kitchen on sneakers with exposed three-inch coiled springs in the heels. Seriously! She says they take the stress off her knees.

The cheerful employees apparently spend most of their paycheck at the tattoo parlor. I notice these things, but then I'm a man from another century.

If you have a sweet tooth, you could get into trouble at the Slice o' Life. Might I recommend the cherry cheesecake, and a rhubarb pie to go? Or maybe you'll chow down on the sugar cookies cut out like little honeybees.

It's early April as I write this, and my bees are just back from California. This was the first time I ever shipped them to the almonds. They're bustin' out of their boxes! I sent 40, got back 38, all but two full of bees and honey. My concerns about the little darlings coming home riddled with mites proved unfounded. I sugar-shake tested all of the returnees. Over half tested zero or only one mite per 300-bee sample. The high count was seven. I used Mite Away (formic acid) Quick Strips between the supers to treat all the colonies with two or more mites per sample. I offset the top super forward an inch to give the treated hives extra ventilation, as this stuff can be hard on queens.

Speaking of queens, next week I expect a shipment from California, followed two weeks later by more queens from a different California breeder. I had pretty good luck with these queens last year. I'm using them primarily to make splits and nucs. So I'm using these queens to start new colonies. I'm not replacing perfectly good old queens with untested new ones, even though the experts say you should. But experience taught me that queen replacement can be risky. Lots of things can go wrong. Far too often, the new queens get superseded. This is a big waste of time and hive energy when you're trying to build up colonies strong enough to fill honey supers.

Maybe someday when I'm a better beekeeper, and my success with new queens improves, I'll try re-queening every year . . . maybe.

June 2014

Courage, and a cool head

I tell people that my beekeeping business is "way more than a hobby, but far short of a livelihood." I love to keep bees, and to make money from them, but I still need a paycheck to survive.

I primarily sell my honey retail, that is, mainly to people I know. I'm a way better salesman than I am a beekeeper.

I sold pollen retail, too, until I sent two people to the hospital with allergic reactions. That's a liability risk that I can't afford. Now I sell wholesale.

Occasionally I do get a call from someone with allergies, who really wants to buy my pollen. So I make exceptions. I put a note in the bag with the pollen. I warn them to try the tiniest sample on the tip of the tongue, before consuming any significant quantity of this stuff.

My handwritten note might not be of much use if I ever got sued, but this is the way I do it. If folks insist, I'll sell to them.

The other day a woman called and said she wanted ten pounds for her brother in Hawaii. She said she'd purchased pollen from me before, even that she'd been to my house. I have no recollection of this. But then at my age, there are a lot of things I don't remember.

It struck me that there are all kinds of restrictions on plant and animal importations into Hawaii, and for good reason. Until recently the islands were a safe haven from Varroa mites.

I never mentioned this to my customer. As long as she didn't kill herself with it, I didn't consider what this woman did with my pollen to be any of my business. Plus this pollen had been in the freezer for at least a week. If there were a bug in there, I guarantee it would be a dead one. And I can always use an extra hundred bucks, so I didn't worry about it.

I know beekeepers who aren't afraid to charge a lot for their honey, and maybe that works for them. But I sell mainly to repeat customers who eat my honey themselves. I charge a little more than the grocery store, and folks still think my honey is a great deal. My market would probably support a modest price increase, but in a recession, I just can't do it. Nobody around here's flush like they were a couple of years ago.

Take the girls down at the bank. They get together and call me. They receive a discount for an eight or 10-quart order, and the money's waiting for me in a little envelope when I make my delivery. They're mainly tellers, so they don't make any addition mistakes. But my guess is that they don't make a lot of money. They have families. Price matters.

My honey reps get my best price. The reps buy by the case and then re-sell. One of the hospital paramedics sells an unbelievable quantity of honey. He's the "honey guy" at work, and sometimes it's all I can do to keep up with his orders.

I sell 8-ounce glass hex jars to a few high-end outlets in Aspen, but they don't exactly fly off the shelf, and sometimes they granulate there. I take them back when they do. This is a big pain. I like to move product. I like it when the phone rings, and the caller needs a case of quarts.

The other day I ran into a customer. He's no spring chicken. I work on Aspen Mountain, and we rode the gondola together. He was going for a hike on Richmond Ridge, and he needed a quart of honey.

He and his wife are old-time Aspenites. She grew up the hottest ski racer in town and for many years worked on the ski patrol. Now her knees are shot. She inherited the family house on Shady Lane. He has Parkinson's, and some days are better than others.

He said he'd just finished a book – Fruitless Fall, The Collapse of the Honey Bee and the Coming Agricultural Crisis, by Rowan Jacobsen. He said it came from a "Buddhist perspective." He said they had it at the library, and he thought I might enjoy it.

A couple of days later, I called to get his mailing address. I wanted to send him a clipping from the National Catholic Reporter about a peace-activist priest with Parkinson's who nevertheless carried on his work for many years.

At the end of the conversation, I inquired about his feisty and charismatic wife. I know her better than I know him.

He said the day before she'd been diagnosed with esophageal cancer. She was outside on the porch talking to a friend. "We don't know what we're going to do," he said. "But it's time to start thinking about crossing over to the other side."

"We're all on that journey," I said.

When I sent them a card, I got one back right away – from both of them. She sounded chipper. He wrote, "Worry is simply a misuse of the imagination."

You have to admire courage, and a cool head. And maybe I should read that book.

October 10

Diving for cover

On a sun-soaked November day down by the Colorado River, Paul and the boys stacked four-way pallets of bees three-high. Semis rolled in three and four times a week to haul the little darlings to the land of almonds and honey. Derrick and his crew pamper them through the California winter with pollen patties, sweet syrup and oxalic acid. In February they move them into the almonds. By late March, they're home again, cheerfully bustin' out of their boxes. In a good year.

I piggybacked most of my best colonies onto one of Paul's loads. I had 'em jacked up on pollen substitute, and on the day I'm going to tell you about, I was getting them all sorted and ready to go.

About 11 a.m. the landowner came by on a four-wheeler with some hunters in tow. He asked Paul how long we were going to be, because his hunters wanted to shoot at some elk bedded down by the river. We'd be in the background.

Paul said we'd be finished by noon. Well, I wasn't going to be finished by noon, but Paul assured me that they were just going to take a few shots, and then it would be over. "You'll be fine after one o'clock," he counseled.

I ate lunch up the river under a mighty ancient cottonwood, then stretched out and tumbled into dreamland on the banks of the Colorado. I awoke to a rainbow of sunlight shimmering on a gentle riffle. I found this magical moment so satisfying that I went back to sleep.

A little before one, I heard repeated gunfire from downriver. "Let 'em have their sport," I mused. "They'll move on, and I can go back to work."

By 1:30 I was back on the job, pulling Apivar mite strips – slow going, because they can be hard to find suspended between the frames. You sometimes have to smoke the bees off the top bars to see the strips.

I was under the gun, because I was behind schedule, which for me is pretty typical. Paul was coming back the next day to do the final hive stacking. The truck schedule kept changing, and he wanted to be ready.

An hour later, I was still pulling strips, when I heard shots again. Across the river two or three hundred yards away, someone in blaze orange was shooting at elk running along the riverbank. I was directly behind the elk.

I hit the deck and sprawled behind four-way pallets of bees. Very undignified! The firepower was impressive. It sounded like the Battle for Mosul. I have no idea how many animals got shot or how many hunters were shooting. I waited for a lull, then took a peek. I saw more

---

hunters but no dead animals. Could they have missed that many shots? Then I heard another volley. Were they warning me to get the hell out of there? The whole situation felt creepy, not to mention unsafe, so I closed up the hive I'd been working on, grabbed my smoker and hit the road, leaving gear scattered everywhere. You could call it an inglorious retreat.

Things move pretty quickly at the holding yard. Itineraries change. Trucks roll in. Bees get loaded and trucks roll out. The next day I never had time to combine a handful of colonies that had fewer than Derrick's requisite ten frames of bees for the almonds. You do what you can. In the bee world, nothing ever goes quite as planned. I'll pay the shipping to send all these bees to California, and once they arrive, Derrick can combine the ones that need it. That's just the way it is. As my sidekick Marilyn likes to say, "Don't let perfection be the enemy of progress!"

I always keep some bees back to winter in Colorado. I don't like to put all my eggs in the same basket. I have apricot pollination customers near Grand Junction. They need bees in March, normally before mine return from California. I make less on these contracts than I would shipping those bees to California, but I enjoy the ritual of taking them into the orchards and watching them work their magic in the fruit blossoms. My growers are grateful to have a reliable supplier of pollinators. I feel it's a public service.

Did I ever tell you how I combine hives in the fall? I'll tell you again. First I dump all the bees from one weak hive into a single brood super and put two pollen patties on top. Next, I dump all the bees from another weak hive into a single brood super, and put that super on top of the first hive. I don't put newspaper between the two colonies -- only the two pollen patties. I let the queens duke it out, or not. This technique generally works. At least the little darlings have some extra warm bodies to huddle with through cruel December.

It's nearly mid-November as I write. My bees arrived in sunny California, as far as I know. They're in Derrick's hands now, and God's. The home bees are well fed. I have a little bee yard puttering to do, but I'm basically ready to take a break until February. Maybe I can catch my breath.

January 2018

Double dipping

One great thing about working on the ski patrol is that I can double dip. Aspen Mountain employees and savvy locals know there's honey in my cubby at the mountaintop patrol headquarters.

One fine spring day, I was at the bottom of the hill, where I keep my main honey cache. It occurred to me that the honey cupboard on top was bare. I didn't have a backpack, but I grabbed three glass quarts, anyway. I tucked one inside the front of my patrol vest and cradled one under each arm. I'd done this before, and with this kind of advertising, sometimes I can make a sale to a lift operator on the way up.

It's only three short lift rides to the top, and it was a slow day. What could go wrong?

As I was about to get off the second chair, I spied a young woman struggling to ski down, so I skied down to offer help. I explained that this is an expert ski mountain. There's no easy way down. When I suggested a sled ride to the bottom, she conceded that that might be a good idea.

I made a radio call, and when Ricky arrived with the sled, I took the handles. "Hey, would you mind taking two honey jars to the top?" I said. "I'll buy you a beer later."

I still had the one jar in my vest.

I tied the young woman's skis onto the sled and told her to sit on top and hold on with both hands.

She was just as terrified of her sled ride as she was of skiing down, even though I always thought I was a pretty good driver. I don't know how I could have given her a smoother ride, but she whimpered the whole way.

En route, my radio squawked that someone had hit a power transformer on the same run I was descending. I said I'd check it out.

I stopped maybe 20 feet above the transformer. It sat adjacent to a rail fence surrounding a deck on a former on-mountain restaurant. Now a ski hill maintenance man lived in the building with, significantly, his dog.

But there was nobody around. I radioed back that this was an apparent false alarm. When I finished talking, I heard a faint voice. "Over here!"

I skied right to the transformer, and lo and behold, there was a seven-foot sun-melt hole in front of it. Both the transformer and the hole were marked by a line of bamboo poles uphill of them. Inside the hole, reclining at a 60-degree angle, was a man holding his arm. One of his detached skis was stuck in the fence above the hole.

He said he hit his elbow and that he had no feeling in his hand.

I immediately jumped into the hole with him and radioed for help. The young woman who had been riding on my sled peered down into the hole at us and said, "I think I can make it down on my own now."

My patient was pretty good natured about his predicament. Not everybody would be.

As I splinted his arm and put it in a sling, I detected an unmistakable odor. I said, "Do you realize we're at the bottom of a hole full of dog doo?"

"I hadn't noticed," he said with a wry smile.

After help arrived, I told him he'd have to kick steps and climb out of the hole, although we'd help. As he struggled to get out, I yelled to another patroller to "Grab his good arm!" But she just stood there.

Afterwards, I asked her, "Why didn't you pull on his arm!?"

"It had dog doo on it," she said.

After we had our patient loaded into the sled, I leaned over to secure a strap, and that quart of honey squirted out of my vest pocket and began rolling and tumbling down the steep, icy slope at an alarming rate of speed!

"Uh-oh!" I exclaimed. In my mind's eye, I saw that three-pound glass jar exploding, or worse yet, bouncing and rolling all the way into town.

Max, who is young, and skis better than I ever did, already had his skis on. "I'll get it!" he cried, and in a flash he swooped below the jar and fielded it like a hot grounder.

Of course the whole adventure became the talk of the patrol room. This story had it all – the terrified sled rider, the patient I couldn't find, the ski stuck in the fence, the numb hand, the dog droppings, the maintenance man and his dog, the patroller who wouldn't touch the patient. The honey jar was just the icing on the cake of a story with an ultimately happy ending – the injured party's elbow wasn't seriously injured – apparently he only took an enormous whack to his funny bone.

The other good ending to the story is that my boss has a sense of humor. He had every right to read me the riot act for letting that honey jar take off down the ski slope, and he certainly has no obligation to let me use the ski hill as a honey outlet. But all he did was chuckle.

I still have my job, and you can still buy honey at the patrol room on top of the mountain. If you don't have a backpack, no worries! Surely you have a big pocket in your jacket. You'll be all right, even if you take a spill on your run down, because now I pack my Aspen Mountain honey in plastic.

July 2010

Gettin' 'er done!

Of a dreamy summer's afternoon I labored 'midst my little darlings. Here at my remote Colorado Flat Tops apiary, on private property, behind a locked gate, an hour from home, I felt like I might be the last person on Earth. So I was surprised when my gal Marilyn came roaring up in her little Saturn.

"Some truck driver named Johnny says he's up in Aspen picking up shipments, and he's coming down our way to get your honeybee pollen – like in three or four hours. I can't get cell service up here. You'd better get home!"

"No way!" I said. "The pollen isn't even packed for shipment. I can't do it. Not on four hours notice. What are they thinking?"

"You can do it! Just head home and get on it!"

Marilyn chuckled. "When I told him you don't have a cell phone, he wasn't pleased. He said, 'Lady, this is a modern operation!'"

Back in civilization, I called Johnny. After a few rings, a voice boomed into the receiver – "Yeah!"

I explained my plight but ventured that I might be able to get my 700 pounds of product boxed and ready for shipment that afternoon. I still needed to buy tape. I had boxes to assemble, pollen to weigh. But I knew my new buyer was keen to take possession.

First Johnny wanted to know if the load would be "palletized."

When I asked what size pallet he required, he shouted, "What size? Standard shipping pallet! Forty by 48! I've got 17 on board. I suppose I could give you one."

My hearing is not so great, but he was so loud, it sounded like he was inside my ear.

"You got shrink wrap, right?" he continued.

I said, "Johnny, this is not an industrial outlet. I'm just a guy with some pollen in his freezer. We're talking about fewer than 20 boxes."

Then the kicker: I said, "I'll give you directions to my place. What time do you think you'll be here?"

He said, "Sir, this here's a tractor trailer, and we don't assume no liability. We don't go down nobody's drive. Company policy. And I'm on a schedule. I gotta pick up peaches in Palisade."

"How about the entrance to McDonald's in New Castle?" I said. "Semis park there all the time. You could call me when you leave Glenwood."

"Yes sir!" he screamed into my ear. We had a rendezvous.

The post office didn't have packing tape for sale, but the grocery store did. The cardboard shipping boxes went together pretty quickly. Then they didn't seem quite sturdy enough, so I taped them again.

I had the pollen stored in 3 mil plastic bags, which were supposed to fit inside the boxes, and they did, but the weak link was the bags. I had some blowouts until I figured out I needed to impose a 40-pound limit. So I scooped from one bag to another, until none was too full. I tried to keep meticulous records, but of course I forgot to record one box, and I wound up searching and counting, again and again.

All this time the clock was running. I thought, "If I don't finish by the time Johnny calls, I can always send the leftovers with the next shipment." I asked myself if I might be crazy trying to get this done, because when I hurry, I always seem to regret it. But he was running an hour late when he called, and I had my flatbed packed and ready.

I was parked in front of McDonald's when Johnny showed up. He might have been 50 or 55, all tattooed up with cartoon characters like Donald Duck and Woody Woodpecker. His T-shirt had a big ol' horned toad riding a motorcycle on it. Johnny was a talker. And he was in a hurry.

His monster-tattooed partner was more the sullen one, and mostly he was in the way. Whenever he tried to help, it was wrong, according to Johnny. My shipment got loaded in no time.

Johnny said, "Long day. There's nuthin' in this sorry excuse for a town. We're gonna find us a motel in Rifle. He flashed me a rotten-toothed leer. I might want . . .

And here, gentle reader, I must spare you. Children could read this! But let your imagination run. Yes, that's exactly what he said!

"You might have to go Grand Junction for that," I laughed.

"Junction? That town's full of drug addicts!" This was news to me. I guess it depends on the crowd you run with.

Finally I said, "I thought you were in a big hurry. You said you had peaches to load in Palisade!"

"Oh, that," he said. "That's tomorrow afternoon. We're gonna party tonight and sleep in tomorrow."

As they pulled away, I sighed and counted my blessings. Rarely do life's adventures unfold as planned. But we got 'er done!

November 2013

Author in the bee yard. © photograph Irene Owsley.

Grand Slam

Yesterday was a grand slam. First I got a massage from a chiropractor of rare talent. She trades her services for our honey, beef and lamb.

Then, a couple of hours skiing on Aspen Mountain. The Aspen Skiing Company kindly awarded my gal Marilyn and me lifetime passes, in recognition of my 44 years on the ski patrol. As a patrol retiree, I no longer have to tackle the gnarliest runs or carry bundles of bamboo on my shoulder while skiing. I don't have to listen to my patrol radio or haul a toboggan. Skiing's fun again!

Next, my annual Medicare physical, from a physician who sees things my way. Look, I'm 70 years old. Nobody lives forever; I'll take my chances. Let's not go chasing test results that might sit me bolt upright in the middle of the night. And please, no risky biopsies in certain sensitive bodily nether regions. I've been in the hospital. I'd rather spend my time in the bee yard. The good doctor took all this in and nodded sagely.

I told him that cannabidiol, or CBD, helped the arthritis in my hands. Used to be I couldn't button my shirt without help. CBD might not be legal where you live, but here in wild west Colorado all the pot shops stock it. CBD is derived from hemp but does not make you high.

Then, after my physical, I walked across town to the annual mid-season Aspen Mountain Employee Appreciation party, of which I was a sponsor. Lesson learned: Next year we should order more pizza!

I got caught up in the endless complications of submitting to Colorado's state pesticide czar recommendations for amending the state's Managed Pollinator Protection Plan (MP3). The state's current plan basically states that pesticide applicators ought to follow label directions, if it's convenient. The Colorado State Beekeepers Association (CSBA) argues for more stringent guidelines.

Michele Colopy from the Pollinator Stewardship Council did the heavy lifting to get the CSBA's MP3 recommendations to CSBA members so they could sign them online. There were details, questions and last minute changes. I needed to stick around on a day I'd planned to ski, so the skiing got canceled. I wanted to ski in the morning and look at bees in the afternoon on the way home.

I still checked the aforementioned bees at 1:30 sharp, because I'd made a reservation the week before to do so. They belong to a billionaire who likes honey bees and pays me to look after his. You could Google this gentleman's name. He's controversial.

On most properties where I keep bees, I come and go as I please. At this particular apiary, however, my instructions are to call before I visit the bees, so that there's someone to "buzz me in" through a series of locked gates, and so they know I'm coming.

Last June I sold this guy four hives, for a song. Mea culpa, but he pays me to take care of them, so everything's all right. It's just that beekeeping is a little more complicated when it's on a billionaire's estate. Take hive access: I can only visit his little darlings Monday through Friday, nine to five, provided I've called ahead. I can't just drop by if I happen to be in the neighborhood.

I don't deal with the billionaire directly but rather through his personal assistant. The assistant is a good kid. He relays information about the bees to the owner, who is rarely mentioned by name. He's simply "the owner." The owner knows a little about bees and has concerns about their care and feeding. I'm working to win his confidence.

Then there's the owner's chef. He got indignant when the honey I dropped off last September crystallized. I tried to explain, but there are people you just can't get through to.

The hives themselves are bomber. When I sold them last June, they were just four strong colonies looking for a square meal. In September I pulled over 200 pounds of honey, which around here is a decent harvest. Afterward, the bees got on a short but wicked honey flow, so the hives went into the winter plugged. When I visited them last week, on February 7, all four had at least ten frames of bees. They were still dead heavy. I wish my own overwintering colonies looked this good!

You don't want to gush or brag, because honey bees can surprise and disappoint you. So when the owner's assistant asked me what update he could provide the owner, I simply said, "They look fine."

Marilyn met me at the employee appreciation party in Aspen. She can hold up her end of a conversation. She's very popular. We chatted with a woman who took beekeeping lessons from a "bee whisperer." The woman's an old friend, and I didn't lecture her about going off the deep end of bee husbandry. Maybe I can bring her to her senses yet. I'll make sure she gets an invitation to come to the June 9 CSBA meeting. Our keynote speaker is Sam Ramsey, who can tell you things you never dreamed of about Varroa mites.

At the witching hour of 7:30, I threw my arm around Marilyn as we walked to the bus stop. On the ride home, she rested her head on my shoulder and closed her eyes. We were back at the farm by 9. The blue heeler Pepper greeted us at the door. It had been a big day, but it wasn't over yet. Looking back, I'd call it a grand slam.

April 2018

Keeping 'em alive

"Why don't you just send all your bees to California for the almonds," Marilyn wondered. "Then you'd have the winter off." Maybe my gal was onto something. Just get on the gravy train.

I only send the cream of the crop. Ten frames of bees minimum, in November -- Paul's rule. He ships truckload after truckload. I tag along. This is the second year I've done this, the second year I sent 40. Last March, 38 came back, nearly every one begging to be split.

This year I could have sent more, but life's little complications got in the way.

My bees that wintered here in Colorado didn't fare so well. They never do. And it's not yet March, the cruelest month. March, the winnowing time, when the strong prosper and the weak perish. How is it that bees make it through December and January, only to dwindle and vanish in March?

So Marilyn's idea intrigued me. Combine weak colonies to make strong ones, and send them all to the Promised Land for the winter. Let Derrick take care of them out in the Central Valley or wherever it is they go out there. Then make splits when they come home. And cash my pollination check. Don't forget that.

The home bees get so needy. I put them on pollen supplement in late January. They can't get enough of it. They powered through honey stores this balmy winter. A measly thirty-some colonies keep me busy, and broke.

You would have thought they'd be relatively mite-free after the grab bag of treatment tricks I threw at them through the fall and winter: thymol, Amitraz, formic acid, oxalic acid. Colonies got different treatments, but that apparently didn't matter. I tested four hives last week. Every one has Varroa.

Do you think almond pollination is a bad idea? Do you believe "they come back riddled with mites," as I once told a reporter? Look, if I send my bees to the almonds, they come back with mites. If I leave them here, they get mites.

Maybe I should go with them to California. Maybe I could make myself useful and not just be in the way. When I dream of the almonds, I can't get Judy Collins out of my head. She croons her haunting ballad about leaving home to follow her rodeo cowboy. "Someday soon, goin' with him, someday soon . . ."

The nasty good California rockabilly band Cracker played an outdoor Valentine concert in Aspen last night, right at the base of the ski hill. Marilyn and I swayed to the music, danced on the snow. I still had my ski patrol pants on. She pulled on my heartstrings when she looked up at me with shining eyes. She's such a sucker for Valentine's Day. And it tugged on my heartstrings again when Cracker sang a sweet song of longing with the line "back to the almond groves." It made me think of my little darlings in the land of milk and honey. Someday . . .

We beekeepers have it made. As the world around us spins into chaos, all we have to do is keep our bees alive. That's it! Got bees? Lucky you! Honey prices hover just under the stratosphere. Bees equal honey equals money, and the world's a better place. You could be rich, especially when the almond growers write the big checks. And the more bees you own . . . All you need to do is make some splits. So why is keeping bees alive so hard?

Last May I had 120-plus colonies. Now I count maybe 75. Where did they go? European foul brood and chalk brood cripple my colonies. American foul brood rears its ugly head. I run out of replacement queens. Queens die, or won't put out, and I throw those hives onto good ones. Starvation sneaks up on me. Relax, and Varroa mites eat my bees alive. My operation slowly dies from a thousand cuts.

The world tips its hat to the beekeeper. Our task is so noble, so daunting. How do we do it? "Isn't it true the bees are dying?" everyone asks. I never know what to say. I know that I somehow keep going. When I have a bad year, I buy more bees from a better beekeeper. You can always tell the better beekeeper. He's the one selling, not buying.

Is beekeeping sustainable? I'm not even sure what "sustainable" means. Sustainable for me, in my lifetime? For 100 years? For eternity? I only know that, like honeybees, some beekeepers thrive, while others flounder. In the teeth of persistent and ominous drought, successful beekeepers send their bees to the almonds, and they bring them back stronger than they sent them.

I know that you can learn from the best and the brightest, but you need to figure things out on your own, too. I know that bad advice is as plentiful as good. There are a thousand ways. Do you have a better one? I hope you do! Did you read about it in a book? Did it come to you in a dream? Wonderful! But be wary. Dare to doubt. Honeybees will confound you every time. Beliefs you hold dear need testing in the white heat of the bee yard. The test is simple: Can you keep your bees alive?

April 2015

Life is a gift

This morning on my way to Garfield Creek with a load of pollen traps, I did a double take when I saw the hospital ambulance parked in front of Patti's Main Street Coffee House. The paramedics like to stop in for breakfast. One is a bilingual Latina with flashing dark eyes, and I hoped I might strike up a conversation while I filled my coffee thermos.

The paramedics were sitting by the coffee urn, where I'd hoped they'd be, but this time two men were wearing the uniform. I used my pickup line anyway and cheerfully inquired, "How's business?"

"It's good, or it's not so good, depending on how you look at it," one of them said. "Without other people's misfortunes, I wouldn't have a job. But today so far it's quiet."

He looked vaguely familiar.

From behind the counter Patti asked what I was up to, and when I mentioned, "bees," the talkative paramedic stood up.

"You're the beekeeper who got buried in that avalanche on Aspen Mountain!" he exclaimed. "I drove you to the hospital! You were buried for nearly five minutes. You were blue when your ski patrol buddies dug you out! Irreversible brain damage starts at six minutes without oxygen! You were so lucky!"

What a memory! "Maybe I did go too long without air," I said.

"This could explain a lot," Patti laughed.

Then the paramedic threw his arms around me and said, "It's good to see you again!"

Wow. In the five years since this most unforgettable incident, I'd never encountered anyone so impressed with my good fortune. And despite my initial embarrassment at his display of emotion, I really was touched.

Now Patti wanted to hear the whole story. I generally don't go blabbing it around, so gentle reader, I'll leave the details to your fertile imagination.

But I will tell you the lesson learned from a careless mistake, and it's an important one, so don't you ever forget it: Simply put, life is a gift.

My gentlewoman rancher first wife Cathy just bought 65 acres of Flathead Lake, Montana shoreline. She planted it in sweet clover for soil improvement.

I said, "Sixty-five acres of clover! You're going to make some beekeeper very happy!"

"Really?" she said. "I never thought of that."

I haven't always kept bees, so how would she know? I might obsess about them, but why should most people give bees a second thought?

When people say to me, "So how are the bees?" I always say, "The survivors are doing great!" Then I tell them way more than they really wanted to know.

Our western Colorado winter was relentlessly cold, and the little darlings never got out of the hive for a couple of months, beginning in early December. When I checked the first week in February, there were some signs of dysentery, but all in all they looked pretty good. I'd lost two of 75. Then the weather got nice, and the weaker colonies dwindled, until by April I'd lost a third.

I was feeding pollen supplement in the spring, so I watched those hives die right before my eyes. Every week I'd lose a few. I'd shrug and tell myself the worst was over, but it was a long time before it was.

I took all of my survivors to Grand Junction and Palisade for the orchard bloom, where they prospered. Despite an extraordinary chalk brood epidemic, by mid-April I had plenty of colonies to split.

I get most of my queens from Paul, who purchases them by the hundreds if not thousands. He buys from a handful of producers. He knows the commercial beekeeper secret handshake and always gets the inside word, so I get some pretty decent queens.

Still, introducing even good queens is never a slam dunk.

You hear this stuff about re-queening every year, but why would you do that? You could kill your best producer and replace her with a dud. It happens all the time. Sometimes a hive won't accept a new queen, or if they do, they supersede her right away. New queens can be poorly mated or even drone layers.

I mostly use my new queens for making splits and nucs.

This year I had some new queens that I couldn't use right away, so I banked them in a queenless nuc. That's the way you're supposed to do it. They were in plastic cages inserted into a plastic JZ-BZ shipping bar. I just squeezed the bar between two frames in the nuc.

Except the nuc turned out to have a queen, after all. Didn't matter. I held a dozen queens for three weeks, and only lost one.

The dumbest thing I ever did with queens was leave a box of them on the kitchen table for a weekend while I re-kindled an old romance. I never even gave them water. When I got back, half were dead, and the rest didn't look so great. That's how we learn, however, or at least how I do.

Marla Spivak from the University of Minnesota spoke at the June Colorado bee meeting at Paul's place in Silt. She did a little queen grafting demonstration, and whenever somebody came up with a question she didn't know the answer to, or maybe she hadn't thought about before, she'd say, "I'll get one of my grad students to do a project on that!"

It was old home week for Marla, who used to work for Paul. I like it that the beekeeping world really isn't so big.

I'm always a little wistful when the meeting breaks up, and we part from colleagues we likely won't see again until the winter meeting. Because over time, we become more than just fellow beekeepers. We become friends.

Ah, life's a gift.

August 2010

Author with that little darling Katie Lee. Photograph courtesy of Marilyn Gleason.

Liquid nitrogen ice cream

Liquid nitrogen is very cold -- between minus 320 degrees and minus 346 Fahrenheit. And it's not cheap, but Katie Lee told me you never have to waste any, because you can always use it to make ice cream!

Katie flew into Grand Junction, Colorado, where I picked her up. After breakfast we stopped off at a liquid nitrogen filling station. She unzipped her suitcase and pulled out an insulated aluminum 30-liter vessel called a "Dewar." Katie said it got some attention from the TSA folks.

The cheerful Airgas attendant said he'd seen a lot of things but never a Dewar packed into a suitcase. He filled it with liquid nitrogen, and I tied it onto the bed of the pickup. Katie warned me beforehand that you don't travel with liquid nitrogen in a passenger car. "What if we got in a wreck?" she said.

Katie came to the Colorado State Beekeepers Association (CSBA) bee college from the University of Minnesota to teach us about hygienic bees. Hygienic honeybees remove Varroa mites from brood cells and from each other, so this is a very good trait for your bees to have! They also remove from the hive diseased material, including chalk brood and American Foulbrood (AFB) deposits.

On the way back to the farm in New Castle, we stopped off at Paul's bee yard. Katie and I opened up two hives, and from each we removed a frame of sealed brood. Into this solid or nearly solid brood patch we inserted the end of a four-inch-long chunk of three-inch-diameter plastic PVC pipe. We stuck the PVC pipe down into the brood, twisting and pushing it all the way against the plastic foundation. We then laid the frame horizontally on top of a hive lid, with the PVC pipe sticking up out of it.

Next we poured a little liquid nitrogen into the pipe, to check for leaks. Finding no leaks, we proceeded to fill the pipe with nitrogen. In a few minutes the liquid nitrogen in the pipe boiled off, leaving behind a three-inch circle of frozen dead brood. Done for the day! Now we had to wait 24 hours to see how efficiently the bees cleaned out the dead brood.

When we came back the following day, with the entire bee college in tow, we examined these same frames of brood. In one, the bees had removed virtually all of the dead brood, indicating excellent hygienic behavior. In the other, only half the dead brood had been removed. Not so good!

After Katie showed the results of the test to the bee college participants, she demonstrated how she froze the brood to set up the test. So now we had two more frames with frozen brood. She and I came back the following day to see what we might learn. Results: two more frames of dead brood 100 percent or nearly 100 percent cleaned out, indicating excellent hygienic traits.

We were on a roll! Next we did tests on two hives at Colby Farm. Results: two more hygienic hives. These bees came from a California breeder who each year sells me queens that are prolific and hardy, but I had never before gathered data that might document their resistance to Varroa mites. Since Katie explained that only about ten percent of all hives carry the hygienic trait, my two tests gave me reason to believe my bee breeder was doing a good job.

To summarize, five of six colonies tested positive for hygienic behavior, including three of four belonging to Paul. Since Paul's drones dominate the landscape in this part of Colorado, and since he buys queens that are purportedly hygienic, we may have something going here.

This is a very big deal. Of the plethora of challenges facing our little darlings, mites are one problem that we beekeepers, individually, can impact. We can't necessarily control our bees' access to good forage, or their exposure to chemicals, but we can and should reduce their stress from Varroa.

Meanwhile, as Katie gave her demonstration to half the attendees, Paul lectured in front of the honey house to the other half. He showed off frames of AFB, European Foulbrood (EFB) and chalk brood. You can't get this kind of education out of a book. With AFB, you have to see it, touch it, smell it. Once you recognize AFB, you don't have to send off samples to some bee lab to confirm your diagnosis. You'll know. I can sometimes smell AFB from ten feet away. You need to be able to identify AFB, and you need to deal with it. There's more than one way. But recognizing it is the first step.

After all this learning under a hot June afternoon sun, we clung to the shade next to the honey house. As Katie made liquid nitrogen ice cream, I sensed there was no way I was going to herd these wilted beekeepers 300 yards across the property to a classroom set up for a CSBA general meeting. So I stood up on a chair right there in the shade and called the meeting to order.

It was a rewarding day, an informative bee college, a spectacular banquet following. My gal Marilyn and I got to have Katie with us at the farm for the weekend. Damn! Life can be good.

August 2017

Katie Lee making liquid nitrogen ice cream. Photograph courtesy of Marilyn Gleason.

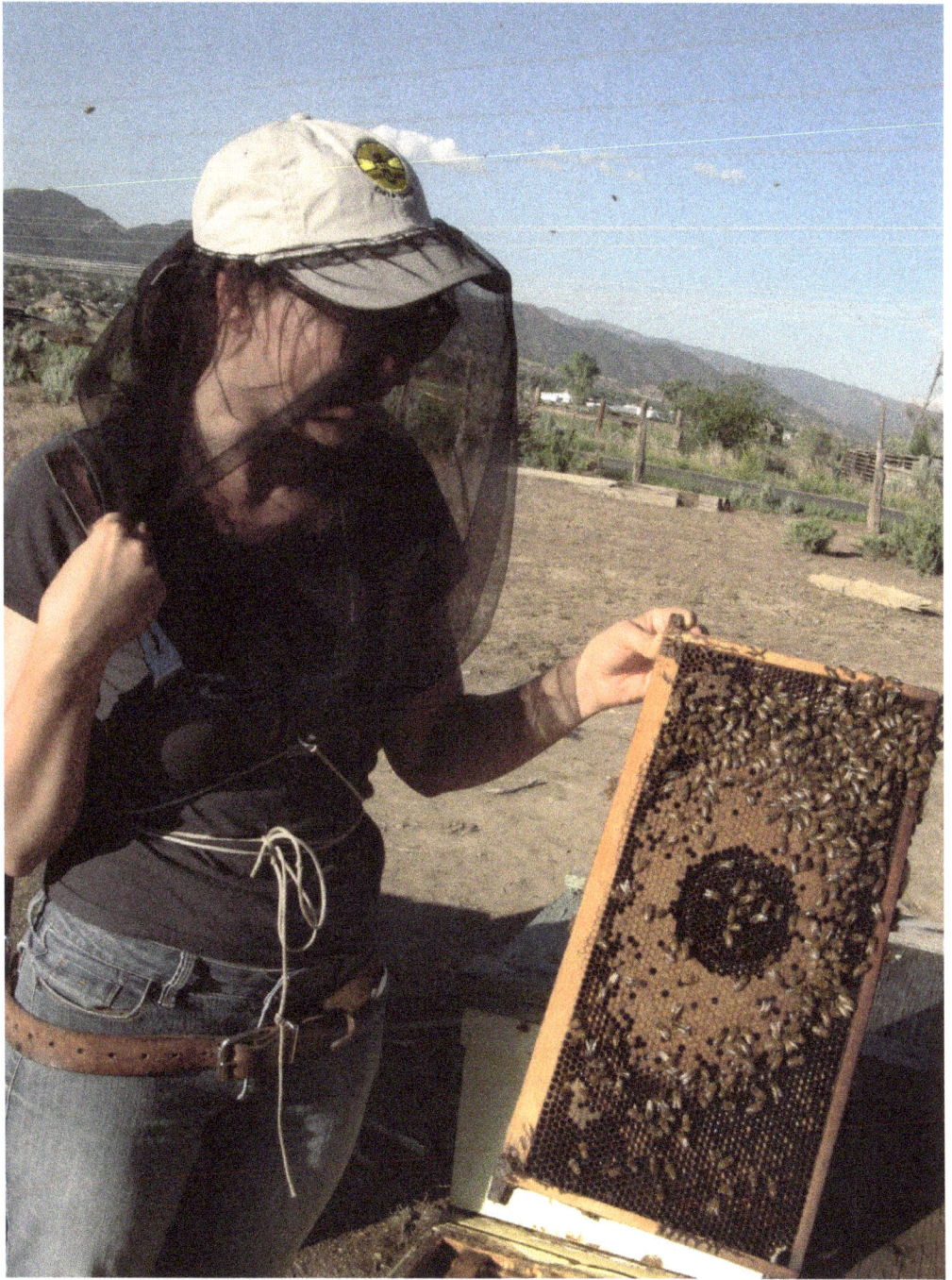

Testing for hygienic queens: Katie Lee shows off a circle of previously frozen brood, cleaned out by hygienic worker bees. Photograph courtesy of Marilyn Gleason.

Marla and mites

I did some dumb things when I was a kid, so I didn't mind helping out a young man who could have wound up in a whole lot of trouble.

Driving down from Garfield Creek with a load of honey supers and thinking life was just peachy, I came around a corner, and slammed on the brakes for a Toyota Tundra pickup lying on its side, nearly blocking the whole road.

The driver was out of the cab. He might have been 20. "Holy guacamole!" I exclaimed. "How'd you do that?"

"I guess I was hot-roddin' it a little," he confessed, "I'd like to get back on my wheels. You don't got a chain, do you?"

"Something better," I said. "I've got a big ol' rope and a 460 under the hood. This'll be easy."

It was late Friday afternoon, but the kid didn't reek of beer. I do suppose he'd had a few.

The rope was a ski patrol rescue relic, 100 feet at least. I doubled it and threw a bowline around the truck's frame on the high end.

"Is that a square knot?" the kid asked.

I didn't want to pull too hard. I didn't know what was going to happen. When the rope went taut, the Toyota teetered, then slammed and bounced on the pavement, right side up! By the time I got out to untie the rope, the kid already had his truck started. "Nice thing about a bowline," I said, "You can untie the damned thing after you put a load on it. I saved you a ticket, you know."

"I owe you big time," he said.

I didn't think so. "Not me, but maybe somebody else," I said.

"Oh, I get it," he said.

Speaking of doing favors for strangers, our own MacArthur Fellow and patron saint of beekeeping Marla Spivak says you extend a kindness to your entire beekeeping community when you control your Varroa mites. Marla hit on this at the Western Apicultural Society meeting this fall in Boulder, Colorado. Especially with the proliferation of backyard beekeepers, too many Varroa-plagued hives go untreated, she opined, and the mite problem has gotten way out of hand.

Marla argued for honeybee "herd immunity" from Varroa, much like the immunity conferred onto a human population when a majority of its members get vaccinated. If most of the people in your home town get vaccinated for say, smallpox, and then suddenly someone comes down with it, the disease is unlikely to unleash a pandemic.

Your hives can be a pit stop on the Varroa transmission highway, or they can be the dead end that saves the bees that live down the road. So gentle reader, maybe it's not just about you and your little darlings. Maybe it's about the rest of us, too.

Marla speaks softly and carefully, never jumping to conclusions. Throughout her distinguished research and university teaching career, she has always promoted bees and beekeeping, ever eschewing easy answers and radical bee ideology. She tirelessly seeks the middle way. We should probably send her to Congress.

Her talk also covered the importance of an ample, rich floral diet for honeybees and native pollinators. Bees that eat lots of pollen produce an abundance of vitellogenin, a blood protein that allows the glands in the heads of young adult bees to secrete good brood food for larvae. Among other benefits, a healthy diet makes it possible for honeybees to detoxify moderate quantities of pesticides, and make brood food largely free from contaminants. Marla makes this all sound miraculous and wonderful, which of course it is.

She approves of the movement to ban neonicitinoid pesticide production and use, because it forces discussion and research on this important issue. Native bees are more profoundly affected by the neonics than honeybees, she notes, because their larvae's detoxification genes are not well developed, and they eat the pollen directly, rather than via nurse bees. But her goal is the judicious use of pesticides, not necessarily their outright ban. "We need to have our pollinators and some of our pesticides, too," she concludes.

She reported on a study tracking a commercial beekeeper's bees over three years, from North Dakota to California. The bees with the best forage had the least winter kill. Remarkably, these bees also showed the greatest exposure to pesticides.

The next morning our Boulder hosts Beth and Dave asked if we might do a sugar-shake mite test on her bees. We stood in her bee yard in the rain, shaking powdered sugar-coated Varroa out of a quart mason jar, like coarse pepper onto fried eggs. We didn't even shake very hard and counted 40 mites in a 300-bee sample. I said, "I guess we've seen enough!" We went ahead and treated her three colonies with Apivar, a time-release amitraz strip.

Beth asked my opinion about hive wrapping for winter insulation. I said, "Look, your bees aren't going to freeze to death. Don't sweat the petty stuff. But do treat your mites. Now that's important."

Her beekeeper neighbors would thank her.

December 2015

Our place in the cosmos

In the wee hours, I turned to the gal Marilyn. "Can't sleep," I confided. "I have too many bees. I don't have enough honey supers. I don't have enough time. I'm still three years behind on taxes. I never should have bought that yard from Paul."

She said, "It's all going to work out, Ed. There was this astrophysicist on the radio today. He reported that the New Horizons space probe took ten years to get to Pluto, at 36,000 miles an hour. At this speed, to travel from one star to another in our galaxy would take a hundred centuries. The universe is really big!

"So interstellar travel isn't so practical at 36,000 miles an hour. But there are these worm holes that are related to black holes. Worm holes suck matter out of one part of the universe and spit it out in another. Slipping into a worm hole might be a way to get really far away.

"In black holes, gravity is so intense and matter is so compressed, they think that time and space switch, so that you can't move through space but you can move through time. However, if you fell into a black hole feet first, your feet would start falling faster than your head, so you would break apart, over and over, until you were infinitesimally smashed up.

"Last February space observers reported that two black holes got sucked into each other's gravitation fields and collided, sending ripples through the fabric of space-time. Einstein predicted this, but we'd never seen it before."

Lying there next to my very own amateur space expert at 3 a.m., I suddenly got goose bumps. "What does it all mean, Dearest?" I queried.

She likes to call me Lalo, a Mexican nickname for Ed. "Lalo, it means our earthly concerns really aren't that important, and that means everything is all right. Don't worry about those honey supers."

This is one way of looking at things. I closed my eyes and slept like a saint, but when our old red rooster crowed, the interstellar magic vanished.

I needed at least 100 supers and frames to go in them. I got some prices for supers, unassembled, assembled, painted, unpainted, pine, cedar, with and without foundation, budget and commercial grades. Don't forget shipping. It wasn't pretty.

Look, I didn't have time to nail together supers, but you do what you have to do. In a way, pre-assembled, pre-painted made sense, but there remained the issue of all that foundation. I needed to integrate it with my stockpile of drawn comb. If you pull three combs out of a super and replace them with foundation, bees on a good honey flow will generally draw that foundation, no problem. So if you already have 200 supers with nine frames of drawn comb in each, and you acquire 100 supers and 900 frames of foundation, you can fill 300 supers with six frames of drawn comb and three frames of foundation. That's the math, and that's the way I like to do it.

Except I didn't have 200 supers. I had 162 left in the barn, and some already had undrawn foundation. I was unhappy with my situation, my pathetically insignificant station in the universe notwithstanding. I kicked myself for poor planning.

There's a way around almost everything. If you don't see it, it's generally because you're trying too hard and missing what's under your nose.

I remembered that Tom in Boulder was downsizing his operation. When I called to inquire if he might have some supers for sale, he said, "About a hundred." With drawn comb. When he shot me a price, I said, "That's a steal."

All I had to do was pick them up in Boulder.

Today Marilyn takes the train to Granby for a family anniversary. The Colorado state bee meeting is in Silt, just down the road, tomorrow. The plan is for me to skip the afternoon bee talks and drive my 1983 one-ton flatbed Ford three hours to Granby to meet up with Marilyn and join in the festivities. Then, the next day, flash my senior National Parks pass, drive over Trail Ridge Road in Rocky Mountain Park, and head into Boulder from the north.

This is not a misprint. We're driving a 33-year-old truck over the Continental Divide to pick up those honey supers.

Yesterday I crawled under the beast and found a length of rotten fuel hose. It broke like a pretzel in my hand. I'm so glad I looked!

I'm not the only one working on vehicles around here. The other day a honey customer said he'd stop by on his Harley. I said, "If you see two legs sticking out from under a red Saturn, that's Marilyn pulling her starter. You can sit in the shade next to her, drink iced tea, and hand her tools."

He said, "Marilyn works on cars? Now that's my kind of woman!"

"Mine, too," I confided. I never told him she knows about worm holes, too.

August 2016

Pepper the pitbull

My sidekick Marilyn's niece ended her career as an extreme skier when she broke both legs jumping off a cliff at a Snowmass competition. We attended her May wedding near Taos. Galena married (of course!) a kayaker and Telluride ski patroller.

Her father rowed her to the ceremony on the banks of the Rio Grande, in a dory.

The "preacher" was a Grand Canyon river guide, the guests professional skiers and boaters. There were sunburned, sandaled women there you wouldn't want to arm wrestle. Everything was laidback and behind schedule.

I didn't tell any of my own Grand Canyon stories, like twice flipping my raft in House Rock rapid -- with different wives! You're not going to impress the pros with that story.

Before we slipped away early under cover of darkness, a dazzling bride confided, "We've got ten acres down by Sawpit. I'd like to get some bees." Ahhhh. Gentle reader, for those who've sucked life's marrow and now yearn to become wise, bees are the new frontier. Today nothing is more intriguing, more green, more hip, than keeping bees.

It wasn't always so. When I started 20 years ago, it merely confirmed what some friends already suspected – that I was nuts!

We'd just gotten back from Taos, and already Marilyn was muttering that our getaway was too short. It was three days! Didn't I twirl her on the dance floor and take her to those hot springs by the Rio Grande she'd been dreaming about? What does she expect in the merry month of May, when the bees run me ragged?

I got some new Sundance pollen traps with side drawers, to use for my four-way-pallet hives. These traps have wooden overhead covers, which keeps out most of the chalk brood, so the pollen typically comes out clean enough to eat for lunch. The traps had been sitting under a tarp for two months. When we got back, I said, "Dearest, I need those traps painted this week."

I pay Marilyn to paint, but sometimes I need to get her started. Then I try not to meddle.

The morning she finished, she announced, "I have to go to Telluride Mountainfilm – today." Have to go? Today? This was the first I'd heard.

I watched her roll down the driveway in her '99 Saturn with three OK tires and one nearly bald. "Can you change a flat?" I queried. "I've got AAA," she called out cheerily, waving out the window. "Well, keep it under 55," I shouted. "She'll be back," I mused, "with stories."

This left me home alone with her blue heeler dog Pepper, and lots to do. I put him in the cab of my pickup and loaded a nuc in the back. Then Pepper yelped and came flying out the window, snapping madly at a bee! I thought, "This is crazy! Why torment the dear boy? Better to leave him home, even if wants to come." Because he hates honeybees, and they hate him back.

I headed for the Silt Mesa yard. I brought empty honey supers, thinking the little darlings might be on a honey flow. But the dandelions had come and gone to seed heads, and these bees made not a dollop of honey. Now they were starving! Blame stormy weather, or an inattentive beekeeper. I had six frames of honey in the truck, so I parceled them out and headed home for more.

When I arrived, I was greeted by a strange dog! He looked a lot like Pepper, but his face was bloated like some pit bull mongrel. He had a wag in his tail. I reached for his throat, and it felt like it might be swollen, too, though I couldn't be sure. His lips felt like big fat rubbery pancakes. I called the vet, but she'd taken off for the Memorial Day weekend. I keep my own allergic-reaction Epi-Pen, and I got it out. Then I decided to not panic. It'd been a couple of hours since poor Pepper got stung, and the danger of anaphylaxis should have passed.

Pepper and I went outside, and he started harassing the geese. This was a good sign! I called my neighbor Howard and said, "You need to take a picture of this!" Pepper growled and snapped at Howard's Australian Shepherd. Hey, I think he'll be all right!

Another hour went by. It was getting late. I put Pepper back in the house and ran back to Silt Mesa to feed the bees. "He'll be fine," I kept telling myself, and of course he was. He still had the pit bull look the next day, but by the morning after, he looked like Pepper again.

He still loves the truck, even though that's where this all started. But he stays away from my little darlings. He just hates 'em.

August 2014

Playing Hooky

In January the gal Marilyn and I took the overnight Amtrak from Colorado to Reno, for the 2018 American Beekeeping Federation conference. But before we got on, the train hit an abandoned automobile outside of Chicago. Then, coming out of the Moffat Tunnel in Colorado, it had to slam on the brakes for a skier on the tracks at the base of the Winter Park ski area.

We boarded six hours late in Grand Junction. Sometimes Amtrak can catch up when they're behind. But they'd replaced the wrecked Amtrak engine with a slower freight locomotive, so we continued to lose time and arrived in Reno seven hours late the next day.

Reno advertises itself as "The Biggest Little City in the World." It's an interesting town. I'll say that. Our well-appointed downtown Airbnb condo sat on the banks of the Truckee River, across the water from the blinding lights of the casinos. In an odd juxtaposition of Reno architectural styles, tawdry 1950s and 1960s motels sprout like weeds among glittering high-rise monoliths. The homeless sleep in vacant lots or wherever they can find shelter.

The place drips with sex. Gambling billboards feature women with big-lipstick smiles, leaning seductively on Hollywood-handsome men at blackjack tables. A sign in the window of the Cal-Neva casino features three very naughty ladies beckoning you to play "beer pong" in the Pleasure Pit. The legal brothels are all out of town.

You can get married in Reno. Wedding chapels flourish everywhere. At the Chapel of the Bells drive-through, you can get hitched in your car.

I wanted to hear Randy Oliver's talk about his adventures with oxalic acid for Varroa mite control, on the opening day of the conference. But instead I fell under the sway of Marilyn and Michigan State bee researcher Meghan Milbrath, and went skiing. Marilyn and I picked up Dr. Milbrath at the conference venue, where she was staying. She kept her head down as she dragged her ski bag past other conference attendees, hoping no one would catch her playing hooky. Let it go, Girl! This is Reno, land of forbidden pleasure!

On the drive up to Mt. Rose, Meghan gave me ideas for how to structure a beginners' bee class, sorely need where Marilyn and I live out in the boonies. I don't have time for teaching classes, but somebody needs to do it! Then it hit me – I was talking bees with a professional bee researcher. This was a business ski trip. I could write off my lift ticket and ski rental!

Meghan and I nearly froze when we skied together in Colorado earlier in the winter. Today we had sunshine and mild temperatures, but the wind howled. Mt. Rose had only a couple of lifts open. It rained the day before. Icicles dripped from pine boughs. We jumped on the lift and talked bees.

We discussed neonicotinoids, the controversial systemic pesticides now very much in vogue. Meghan explained that the neonics don't all act the same in the environment and as stressors on pollinators. We talked about risk assessment and how we make risk-related decisions. Meghan brought up pumping your own gas. Gasoline contains benzene, a known carcinogen. Every time you pump gasoline, you assume some cancer risk. Exposure to benzene is an enhanced risk for people in poor health. But as a society, in general we agree that pumping your own gas is worth the risk.

We talked about fungicides and neonics as synergistic bad juju and about recent research pointing to fungicides as pollinator attractors. Bees may prefer flowers coated with fungicides.

Our lift ride went pretty fast! Meghan and I followed Marilyn down the hill. Marilyn likes to be the leader, and Meghan and I like to follow. Marilyn can be hard to keep up with!

On the lift again, we discussed how difficult it is to prove something scientifically, and how preconceived notions and points of view color our perceptions of reality. We talked about the funding of research by interested parties, specifically pesticide research conducted and paid for by pesticide manufacturers. "But who else has the money?" Meghan queried.

We agreed that neonic seed treatments violate a fundamental principle of Integrated Pest Management. In IPM, you use the lowest dose of insecticide that'll do the job, and then only when you absolutely need it. With neonic seed treatments, you treat the problem when you plant, before you even have a problem. Dr. Milbrath explained that for crops normally coated with neonics, like corn, soybeans, and Canola, farmers can't even buy non-treated seed. They're stuck with treated seed, whether they want it or not. She had to drive from Michigan to Indiana to find non-treated corn seed for a research project.

At the top of the lift, we hiked uphill a few steps to a place where few skiers ventured. We descended through an enchanted forest of wind-battered pines and twisted tree snags. Lake Tahoe shimmered in the distance. Meghan got too far forward on her free-heel skis and augured in face first. She got up laughing, the little darling! We never stopped for lunch.

Back on the lift, we kept congratulating ourselves for ditching the ABF convention. Finally I said, "This has gotta be more fun than Randy Oliver's lecture!"

"I'll tell him you said so," Meghan said. Her smile stretched across the snow-capped Sierras.

March 2018

Quick trip to Boulder

My gal Marilyn thinks we should all do our part to reduce global warming, by using public transportation. So when I announce that I'm driving over the Continental Divide to Boulder for a Colorado State Beekeepers Association (CSBA) board meeting, she says, "Why not take the bus? You could come back on the Amtrak."

Boulder is only a three-hour drive from Colby Farm, but Marilyn's idea intrigues me, especially since our five vehicles are all beaters.

So I catch the 7:25 a.m. "Bustang" out of Glenwood Springs, get off in Golden to visit Marilyn's niece, then hop the new light rail to downtown Denver. From Union Station, the "Flatirons Flyer" bus takes me to Boulder, where Neil and Tina pick me up. We arrive ten minutes early for the 5 p.m. board meeting at Kristina's house.

I confess that CSBA board meetings are not models of efficiency. We sometimes wander from the agenda. I tell jokes that make Tina wince. We've been known to whip a dead horse. Three hours and fifteen minutes into the meeting, I say, "Guys, we gotta wrap this up." We've covered our agenda. I'm pleased with the evening's accomplishments. I think we all are.

I crash on the rollaway in Kristina's basement. When I awaken at 4 a.m., I'm pretty sure I'm not going back to sleep, so just as quiet as a little mouse I slip out the front door, walk down Table Mesa Drive and catch an early bus back to Denver. As I step into cavernous Union Station, I stop in front of the flower shop to admire the image of a pink, two-foot wide, spread-winged honey bee.

The Amtrak is running late. My bright-eyed breakfast server buoys my spirits. When I emerge into the metropolitan bustle, the world looks fresh and new, a better place.

What's all this about partisan politics and a nation divided? As I watch an older gentleman hold the door for four strangers, the pettiness and meanness of humanity melts away. Every woman appears to me angelic, every man a noble-hearted brother. Am I in Heaven?

Outside on the street, a crane lifts a beam onto a building under construction. Above me I can see the operator's hands on the controls as he eases his burden into place. He has an American flag in his side window.

I stop in front of the 25-foot-wide, four-story Hitchings Building on Market St. Its dizzying courses of half-high bricks, stone lentils, ornately carved woodwork, arched windows, and concrete gingerbread hearken to the grandeur of the late Victorian era. Today it's a pot shop.

A young man in a tan suit and Bogart slouch hat strides jauntily down the sidewalk and disappears into a law office. I stop to talk to an old woman about her dog.

In line to board the train, I stand behind a woman of a certain age. Her son asks if she can get on the train by herself. She nods in silence. He kisses her cheek, tells her that he loves her, and walks away. She turns sidelong to watch him go. I feel an ineffable sadness.

We strike up a conversation. She lived in Montana once, on Finley Point, on Flathead Lake's south shore. She worked a summer job in a sweet cherry orchard. She misses the place. "Only five people lived on the point when I was there."

"Why did you ever leave?" I ask.

"I had a husband," she confides. "He wanted the bright lights. I never did."

She smiles when I say, "We don't necessarily marry the right person."

In the observation car, a man asks me why the bees are dying. I do my best to explain.

I want to read my Bee Culture, but how can you ride over the Rocky Mountains on a train and not look out the window? In Gore Canyon I watch a bald eagle enjoy a gory lunch on the ice.

Downstairs in the observation car young people drink beer and talk about sports and music. A woman sings the "Peanut butter jelly with a baseball bat" song and dances around the car, as the others cheer her on. I ask the singer if Marilyn and I could hire her to go on trips with us to entertain us and make us feel young again. I ask if they are friends traveling together. No, they say, they only just met. They want to know why the bees are dying.

When I get off in Glenwood, Marilyn picks me up in the school bus she's shuttling across town. I tell her not to get fired on my account. But she's a rebel. She never listens.

I'm almost home. It's 3 p.m., just 30 hours since I left town. I could have made this trip the easy way, alone in my car. I'd have been home hours ago. But for a hundred reasons, I'm glad I didn't.

February 2019

Spider with a white dot

Down the road at Maud's Restaurant, the pumped-up staff eagerly anticipates this weekend's Colorado Beekeepers' June banquet. Last week at lunch my waitress Mary Ann asked me, "Do I have to dress up like a bee, or can I be a lady bug?"

"No problem," I said. "Any friendly insect's OK." Just so the little darling doesn't come as a spider, natural enemy of honeybees.

After I dropped off a load of bees at the Dodo yard on the Flat Tops last week, I took a break in the cab of the big truck. A bee buzzed against the glass on the inside of the closed passenger side window. Suddenly a hairy, mostly gray spider appeared on the same window, about a foot away from the bee. It was one of those jumpy spiders, like you see in the garden -- three quarters of an inch across, including legs, with a black spot on its back, and in the middle of that black spot, a white dot. I wondered if it might be looking for trouble. I thought of rolling down the window for my struggling apis mellifera, but as this bee tried to get out, another little darling was trying to get in to the cab to pester me. So I could have intervened, but for self-centered reasons decided instead to let God sort this out.

You never saw such a truck as mine. A battered and bruised, chopped-off converted 1983 one-ton Ford Econoline van with an 8 by 20-foot flatbed, it hauls 30 colonies with ease. That 460 engine likes a load. I suppose I could double-stack hives, and haul even more, but I've never tried. Fuel economy: not so great. Hence I don't drive it much anymore. The floor's riddled with holes, and small animals occasionally move in. One year I ran 300 hives up by Steamboat Springs and camped in my bee yards. A little mouse kept me company in my never-tidy truck most of the summer, presumably dining on lunch scraps. So it didn't surprise me when I learned I had a pet spider.

The bee seemed to be losing her grip, slipping down the frame, toward the spider. Meanwhile, the spider hopped around and turned tight little circles on the window, like my gal Marilyn's blue heeler dog Pepper when he's all worked up. I wondered if this would take all day. But no. Suddenly the spider lunged in the direction of the bee, closing the distance between the two to perhaps four inches. Then it stopped. The bee wagged her stinger in front of the spider as she continued to slip down the window frame.

I couldn't take my eyes off the two. Now, as the bee slipped, the spider slowly, ever so slowly, climbed upwards towards her, until the two sat less than an inch apart. It felt like watching Janet Leigh in the shower in the movie Psycho. I knew the story ending, and I did nothing. In a lightning move, the spider attacked. The bee struggled for a couple of minutes. I could clearly see the spider's neon green fangs sunk deep into the honeybee's back. Neon green fangs? There are things in this life, gentle reader, that I could never make up. Why would a gray spider with a white dot have neon green fangs? (The better to bite you with, my dear!)

Do you think less of me now, cold voyeur that I am? Listen, insect life is cheap. I kill more bees than this every time I replace a hive cover. If you fuss with bees and try to save them, you'll never get anything done in the bee yard. Are you squeamish about the crunch of honeybee exoskeletons when you put a super on your colony? Toughen up, Buttercup. Do the best you can and move ahead. Think of the hive, not the individual bee, as the living organism. Otherwise you've got too much death on your hands.

Some tender souls object to the noble craft of keeping bees because we beekeepers kill honeybee queens and replace them. This is crazy talk. Queens die naturally or fail, bringing down entire colonies with them. Our job is to promote the health and vigor of the bee hive organism, not to fret about this bee or that queen. Bees hatch from eggs and live their lives, be they short or long. Then they die, just like us. So stop worrying about it, OK?

I went back to work erecting a solar electric bear fence around my apiary, like I've done a thousand times before. Pound the steel posts, put PVC pipe around them for insulation, hang the woven-wire fence on the PVC with tie-wire, hook up the charger and solar panel to the battery. Wire the charger to the fence and to ground. Turn the charger on. When I hold a blade of grass against the fence and receive a tiny shock, I know I'm finished.

Back at the big truck, the spider and her prey are gone. The spider must have her web somewhere in the cab. I wonder where.

August 2015

Terry Bradshaw

The reason I call her "my gal Marilyn" is because Terry Bradshaw does. She taught him to ski. A few weeks later, on David Letterman, the iconic Hall of Fame quarterback gushed about his vacation in Aspen and about "that gal Marilyn." At one point, Letterman interrupted to say, "Wait a minute! You're married man. Who's this gal Marilyn?!"

Bradshaw's "aw shucks" persona is no put-on, according to my gal. While he was making a cell phone call, Marilyn overheard him say, "How's your mom and them?"

Marilyn cracked the whip on me last October after we got back from the Apimondia bee conference in Ukraine. I landed in the hospital prior to the trip and couldn't do any fall bee work before we left. Now it was mid-October Indian summer here in Colorado, and I hadn't poked my head inside a bee hive for six weeks. Paul pulled my honey for me in September, but I hadn't treated for mites or fed any light hives. I was pretty confident my mite load was low, but I still needed to get 40 hives ready to go to California for the almonds on one of Paul's semi loads. I wanted to get going, because you never know when Old Man Winter's going to show up around here.

Experience taught me long ago that the bees always have to come first. Once upon a time I went to Santa Fe for the weekend. When I got back, half of the queens I'd left on the kitchen table were dead. You never forget something like that.

And it was in Santa Fe that I was scheduled to speak at the Western Apicultural Society (WAS) meeting this fall, hard on the heels of our return from Ukraine. I felt overwhelmed. The talk entailed a day of preparation, plus three days on the road. Right when I had perfect weather to catch up with the bees! I told Marilyn I was backing out. The bees had to come first. But she wouldn't hear it. I'd given my word. I could take care of my little darlings when we got back. Everything would be fine.

We left at the last minute. We always do. Marilyn drove, while I rehearsed my talk out loud. We rolled into Santa Fe after dark.

I said I was scheduled to speak at WAS, but I was really invited to speak on the bus tour on the last day. The bus tour featured a visit to the tiny, historically Hispanic settlements in the rugged mountains on the "high road" between Taos and Santa Fe, highlighted by a visit to Melanie Kirby and Mark Spitzig's Zia Queenbee Farm. This was on the way home for us, so we followed by car. The lunch stop was at the Santuario de Chimayo, an early 19th century church. Pilgrims believe that holy dirt from a hole in the church floor heals the sick. Some believers eat the dirt.

"You should rub some on your bald head." Marilyn said with a wry smile.

At the community theater in Penasco I gave my talk about our adventures in Ukraine. But first I had a little surprise for the audience. I pulled out a frame of new comb riddled with textbook American Foul Brood (AFB). There were puddles of AFB brood goo in the bottoms of cells, so you could do the "ropey test." You dip a matchstick in the goo, and when you pull it out, the goo sticks to it and stretches like mucus. There was dried up scale in the bottoms of cells. The capped brood cells were sunken and shotgun-shot with holes. And it was all easy to see, because it was in new white comb.

The theater lighting was a bit dim, so I handed the frame to someone in the front row and told her she could take it on the bus to pass around. Because you can talk about AFB all you want, but you really have to see it. Then you don't have to send off samples to labs, or wonder. Once you've seen AFB, and smelled it, you don't forget. So I thought my sample might be a rare educational gift for some.

I dived right into my talk, which had nothing to do with AFB. I wasn't very far into it, when a gentleman raised his hand. He suggested that my AFB sample was an inappropriate gift, because by virtue of handling it, folks might transmit the bacteria to beehives.

This struck me as farfetched, like catching AIDS by shaking Magic Johnson's hand. We weren't even looking at bees that day. I brushed off his remarks and kept to my topic, but did I detect a pulse of alarm ripple through the crowd?

Outside in the sunshine after my talk, I saw beekeepers talking animatedly as they examined my AFB frame, but some stood back, as if the frame were radioactive, or maybe harbored Ebola virus.

I thought somebody might want to keep my specimen for reference, but before the bus took off, a woman handed it back to me, wrapped in its original plastic bag. "We didn't know what to do with it," she said. I didn't know what to do with it, either, so I brought it home. It's still wrapped in that plastic bag, in my closet.

Back in Colorado, Indian summer droned on for weeks. Then one stunning blue-sky late November day, I did a final round of sugar syrup feeding. That night it snowed 17 inches. A couple of nights later, the mercury dropped to single digits. Old Man Winter was here.

April 2014

The Good Samaritan

The pot grower down the road told me his bees are doing fabulous. I asked him how his mites were doing. "What mites?" he queried.

I said, "We just had the Colorado State Beekeepers meeting practically in your front yard a couple of weeks ago. You could have learned all about Varroa mites."

"I heard about that meeting," he said, "but we were gone that weekend."

I advertised the hell out of this meeting, and I have other beekeeper neighbors who know nothing about bees but were still too busy to learn when they had a chance.

Well, my Colorado River Valley bees aren't doing fabulous. Today is July 9, and we haven't seen a raindrop since early June. The mercury tops out in the low nineties. We've had wildfires, slurry bombers and helicopter water drops.

Strong hives made a little honey, but I don't have very many of those. I split every hive I could in April and May. The bees mostly missed the dandelions due to stormy weather, and the first cutting of alfalfa yielded little nectar.

Right now the little darlings can't even seem to find pollen. I'm feeding pollen substitute to maintain healthy brood.

Derrick told me he only quit feeding syrup three weeks ago. He got his bees strong enough that they were able to make some honey. Me? I haven't got a full honey super on any of 90 river valley colonies. Maybe there's a lesson here.

Meanwhile, my two high altitude yards continue to plug out honey supers. Diversification is not a bad thing.

At a recent bee meeting, Derrick did a slide show on his strategies for successful commercial beekeeping. He threw in some slides of semi loads of bees stuck axle-deep in the mud during this year's rainy California almond season. Just when the situation looked perfectly hopeless, a guy showed up with a backhoe and tried one way after another to extricate Derrick's mess. As soon as he finally winched the trucks to drier ground, he vanished.

Next day Derrick tried to find his Good Samaritan, but all he had to go on was a first name -- Jose.

When he finally found his man, Derrick couldn't get him to take "a couple of hundred bucks." Jose didn't do it for the money, he explained. He just wanted to help a stranger.

This story should be in the Bible. Maybe it already is.

I went to this meeting of the Colorado Professional Beekeepers Association (CPBA) for a couple of reasons: One, to offer an olive branch from my own Colorado State Beekeepers Association (CSBA). The CPBA recently broke away from the CSBA over philosophical differences, and some bad blood. Two, to carry the gospel that neonicotinoid pesticides are bad for pollinators.

Lyle leads the CPBA, and he's a big man with a big message: Varroa mites are the culprits driving dramatic honeybee losses. Forage loss is a huge problem. Pesticides are a relatively minor nuisance. He's not too fond of backyard beekeepers – or any beekeepers, for that matter – who allow mites to decimate their bees.

He and I share some common ground, and we have some differences.

The featured speaker was Dick Rogers, an entomologist and research manager from Bayer, a company that manufactures neonics. It piqued his interest when I announced that the CSBA joined with the American Beekeeping Federation and the American Honey Producers Association in signing a petition to the EPA to register neonic seed coatings as pesticides.

Arathi Seshadri, a biologist from Colorado State University whose work is supported financially by the CPBA, spoke on controlling American foulbrood within the guidelines of the new animal antibiotic regulations.

Lyle argued that the neonics are not a problem for honeybees, and that neonics in fact reduce the need for spray from chemicals that do harm bees.

I couldn't let this go unchallenged, so I peppered Dick – the Bayer guy -- with questions about research demonstrating that the neonics are particularly hard on native pollinators. Well, he just hadn't seen anything that backed this up. I mentioned a Swedish study on seed-coated canola that showed dramatic decline of bumblebees and solitary bees. He was unfamiliar with that study.

When an industry researcher states that Bayer monitors bumblebees' response to neonics, and everything looks rosy, but he hasn't heard about an important independent research project on the same topic, I start to wonder.

I referenced Marla Spivak's remarks at the 2013 Western Apiculture Society conference, in which she stated that native pollinators tend to be more sensitive to pesticides than honeybees, due at least in part to the pesticide-buffering effects of vitellogenin produced in the heads of nurse honeybees. Arathi the CSU biologist said that I was misinterpreting Marla's remarks.

I started feeling a little picked on, even though I asked for it. At one point I said, "Look, I didn't come here to start a food fight."

After the meeting, Dick and I spoke in private. When I asked about soil contamination from neonic-treated seeds, he said the neonics bind to rocks in the soil and biodegrade over time. He told me he couldn't explain why so much peer-reviewed research cast the neonics in an unfavorable light.

We all want the truth, the whole truth and nothing but the truth, right? So we need to be careful that we don't embrace conclusions that seem reasonable and true, only to work backwards, cherry picking evidence to back those conclusions.

We need to listen to and talk to people with whom we don't agree. How else are we going to bring them to the light?

Later, Lyle and I chatted amiably outside the meeting room on a scorching July afternoon in Salida. I can always learn something from Lyle. A stone's throw away, boaters and floaters frolicked in the Arkansas River. The gal Marilyn tugged on my arm. "Let's go," she said finally. "We need to jump in the river."

September 2017

That redheaded Tina

That redheaded Tina rolled into Colby Farm late from Durango in her '94 Volvo after she had to detour around the mudslides south of Red Mountain Pass. She did stop to look at some bees along the way. My gal Marilyn and I put her in the guesthouse out back. Next morning we all got up with the roosters but lingered over breakfast. Then Tina and I lit out over the mountains for some honeybee politicking.

Tina likes to wear a dress in the bee yard. She argues that it's when bees get caught between tight clothing and your skin that they like to sting you. This inspired me to try wearing hiking sandals when I'm in the bee yard. It works! Bees might pepper my veil and find the holes in my jeans, but they almost never sting my toes. And tickling bees on bare feet lets me know when the little darlings might be thinking about climbing up the inside of my pant leg to check if I'm wearing underwear.

Tina and I are neither of us city folks. I'm generally lost every time I cross the Continental Divide and descend into the Denver/Boulder/Colorado Springs maelstrom. She likes to say, "I don't know the difference between Arvada and Aurora." This resonates, because I don't know the difference either.

To keep the conversation lively as we climbed over Vail Pass, Tina launched into a political tirade. Not bee politics, on which we're staunch allies, but right versus left, with a pinch of The Donald thrown in. We didn't agree on much. After we wore that one out, we started in on religion.

Just when I thought I might have offended her with my take on right, wrong, and the meaning of truth, Tina startled me. "Ed!" she exclaimed. For an instant I took my gaze off the road and looked into the bluest of blue eyes. "I'm really enjoying our conversation!"

Father Bob liked to slip in a joke now and then at Mass. I wish I'd thought to tell this one to Tina:

Pope John Paul is praying in his study, when the phone rings. It's God. He says, "John Paul, my good and faithful servant, I have some good news, and I have some bad news. Which do you want first, the good news or the bad news?"

"Oh, I'll take the good news," the Pope replies.

"I've decided to unite all the world's believers under one religion. No longer will there be religious bitterness or strife."

"That's the answer to my prayers!" John Paul exclaims.

"And now the bad news," God says.

"How can there be any bad news after what you just told me?"

"You're not going to like this," God says, "but I'm calling from Salt Lake City."

This was one of Father Bob's favorite jokes, and it's surely one of mine.

Tina brought a GPS, so she's way ahead of me. I'm still trying to figure out my flip phone. My idea of finding your way in a strange town is to have a human navigator riding shotgun with a Rand McNally road atlas barking out instructions.

We were only 20 minutes late for our meeting, and the polite folks at the Colorado Department of Agriculture were waiting for us. Tina and I were emissaries from the Colorado State Beekeepers Association (CSBA), and the Ag people knew we had a beef. They listened attentively and nodded at appropriate moments as Tina and I did most of the talking. This all had to do with who might best represent Colorado beekeepers on the state's Pesticide Advisory Committee, and who might not.

I'm not going to hang dirty laundry here. Bee politics can damage tender ears. Suffice it to say that there are significant differences of opinion among Colorado beekeepers on just about everything. In the end, the Department of Agriculture officials invited the CSBA to submit its own candidate for the upcoming beekeeper vacancy on the Pesticide Advisory Committee. Great!

But you know how in sci-fi movies somebody figures out that their neighbors are from Outer Space, and they report it to the police chief? Then afterwards they wonder if maybe, just maybe, the chief is an alien, too? That's what it felt like challenging the Ag establishment and the status quo.

It's August as I write, and the drought here in west-central Colorado continues unabated. We have fires all around us. The smoke envelops you. Like a novice smoker, I got used to it and finally stopped coughing. Here in the Colorado River Valley the un-irrigated land feels brittle, it's so dry. Toss your stogie out the car window, and the cheat grass is sure to burn like gasoline. Remarkably, bees in some locations are making honey.

A rancher traded me fishing access for honey. A little creek meanders through his place. In Autumns past I spotted big brown trout that presumably came up from the Colorado River to spawn. I don't know how they'd make it this year due to the low water. I dropped off a case of quarts by the rancher's back door this afternoon. Now I need to break out my rod and see if I can find my fly box. This means temporarily dropping my obsession with bees and mites and honey harvest and taking a day or at least a morning off. I can do this. Really.

October 2018

The Boys

"The widder's good to me, and friendly; but I can't stand them ways . . . she won't let me sleep in the woodshed; I got to wear them blamed clothes that just smothers me, Tom; they don't let any air git through 'em . . . I got to wear shoes all Sunday." – Huckleberry Finn, in Mark Twain's *The Adventures of Tom Sawyer*.

As dusk descends on my honeybee yard on Charles Ryden's Main Elk Creek ranch, I hear the distant laughter and faint shouts of boys running through woods and fields. I contentedly set down my smoker and listen.

Oh, those boys! They embody everything the rest of us lost long ago – openness, innocence, a sense of wonder. They're the luckiest kids on Earth, and they don't even know it.

Cody and Mark live in the old ranch house, and they both love to talk.

When I asked 11-year-old Mark about the fishing in the creek, he allowed as how it was pretty good. I asked him what kind of trout. "Oh, browns, brookies, rainbows," he said.

"How big?" I asked. I wanted to know.

Mark held out his hands and moved them slowly back and forth like an accordionist -- or an honest angler. His hand spread indicated 12 to 18 inches – pretty decent-sized fish for a dinky little creek.

"You sure they're that big?" I asked.

"Oh, you bet," he said.

I had only recently put bees on the ranch, and Mark wanted to know all about the little darlings. But he said he was allergic to bee stings. "Everybody in our family is," he said. This astonished me.

I said, "Wait a minute. What do you mean you're allergic?" People think "swelling up" is an allergic reaction, but it isn't. It's a normal reaction. Not everybody knows this.

He said, "I break out, and I can't breathe. I have to take Benadryl."

"I guess you are allergic," I said.

Just then 10 frisky bulls came stomping, kicking and head-butting across the bridge. I gave a start, but Mark lounged against my pickup just as casual as any rodeo cowboy.

The bulls' rippling muscles made me think of Arnold Schwarzenegger.

Mark knows a greenhorn when he sees one. He said, "I'd move my truck if I were you."

"Why would you do that?" I asked.

"Because those bulls could wreck it if they ran into it. That's what happened to my dad's truck."

Jesus and Sergio and a bunch of the other Mexicans chased after the bulls, whooping and laughing and shouting Mexican cattle calls, as they herded the bulls into a pen. Mark ran to join up.

When he came back, I said, "Those bulls could stomp you. Don't they scare you?"

"Naw," he said.

I said, "Wow, they scare me."

That little philosopher looked at me and said, "Think of it this way. You're scared of bulls. I'm scared of bees. Everybody's scared of something."

Mark doesn't miss a trick. If he had a fence to whitewash, he'd let you help.

Mark's 12-year-old brother Cody – a living, breathing reincarnation of Huckleberry Finn – wears shorts in the summer. No shoes, no socks, no shirt. Just shorts.

One day he said, "Your bees moved into the shed, and I got stung eight times."

"Uh oh," I thought. "This doesn't sound good. They let me keep bees here, and now they've moved into one of the outbuildings and gotten nasty about it. "

"Let's have a look,' I said.

Inside the shed he pointed to a large paper nest by the door.

I said, "Let's get out of here. Those are hornets."

"Oh," he said.

Safely outside, I said, "What'd you do to upset them, anyway?"

He said, "I was inside the shed looking for something, and Mark came over and said, 'Mom needs to talk to you.' Then he slammed the door and ran back to the house. As soon as he slammed the door, those bees tore after me. I went running as fast as I could, but they got me eight times."

"Those were hornets," I corrected. "Not bees." I hate it when honeybees get a bad rap.

In my mind's eye I watched him sprint barefoot down the lane, legs and arms pumping, yelling at the top of his lungs, while bald-faced hornets lit into his bare brown back. Why didn't Mark Twain tell that story?

Youth vanishes like a dream. Rub your eyes, and it's gone. The magic always lies in remembrance, don't you think?

Sadly, by mid-summer Mark and Cody had to wash behind their ears and go back to school. I'm pretty sure they have to wear shoes.

November 2003

Road trip with Paul

Paul and I depart in the teeth of a snowstorm. Over Vail Pass, up the long hill to the Eisenhower Tunnel, no problem, then down to Denver the Mile High City, now south, bound for Texas, where the cotton blooms and blows. We pass semis tipped over by savage winds. We charge ahead, headed for Galveston and the granddaddy of 'em all, the North American Beekeeping Conference.

At midnight we check into an Amarillo motel I wouldn't recommend. I learn how Paul gets up. When the alarm sounds, he leaps. We're on the road a little after six.

In Houston rush hour traffic, Paul misses a turn and heads off in the wrong direction. I use a dog-eared road atlas to guide us back. Paul and I are not smart phone or GPS people. We have no apps. We're from another century.

Early evening in the fog, we park in front of the convention center, surf pounding in our ears. Toto, we're not in Colorado anymore! We register. We run into friends and beekeeping rock stars.

Paul and I walk down a pier to check out the fishing opportunities, then head for my Airbnb at the other end of town. Eight steps up from the yard, just below the front porch, a plaque reads, "High water mark, Hurricane Ike, 2008." Inside, high ceilings, narrow stairways, creaky polished oak floors, a shared bath. Paul has reservations at the Red Roof Inn, but tonight he's on the rollaway in my room. Downstairs an arty poster catches my eye -- smiling topless lady on a Harley, hair streaming in the wind. She's not wearing a helmet.

The North American Beekeeping Convention is a joint meeting of the American Beekeeping Federation (ABF), the American Honey Producers Association (AHPA), and the Canadian Honey Council (CHC). I'd never been to a national bee meeting. I did go to Apimondia, an international meeting, in Ukraine. It was a lot like Galveston – huge tradeshow and an infinity of speakers, some better than others. But in Galveston we heard the best and the brightest, in intelligible English.

At the trade show I got to talk to the manufacturer of my pollen traps. The first batch I bought work just fine, but a later, similar model captures less than half as much. The reps acted interested.

I spoke to vendors of certain mite control products and explained my frustrations.

When I told the owner of a bee supply house that I switched to another company because of shipping delays, he told me, "Just use my name, and you'll go to the head of the line!" Well, that's all very fine for me, but I really didn't tell him that so that I'd get preferential treatment.

I skipped the Flow Hive demonstration, even though I agreed to tend bees this year for a wealthy client who has already purchased four of them. Whenever I think about the Flow Hive, my mind recoils, like it does when I think about nuclear war, or global warming, or my delinquent tax filings.

I especially enjoyed Marla Spivak's presentation on the conversion of the state of Minnesota to the nation's most pollinator-friendly state, with even the governor coming on board.

Former USDA top bee scientist Jeff Pettis gave the low-down on getting kicked downstairs in government when you step on sensitive toes. He got a standing ovation, and right after, when commercial beekeeper Dave Hackenberg gave his Jeff Pettis eulogy, applause shook the rafters.

I learned that, among commercial beekeepers in general, there is plenty of concern about neonicotinoid pesticides. This got my attention, because here in Colorado the commercial guys pooh-pooh neonic dangers. Well, some do. Well-documented losses by commercial beekeepers from other states get dismissed as "PPB," or "piss-poor beekeeping."

At the commercial beekeepers' breakfast a big topic was mites developing resistance to amitraz, the commercial go-to miticide. As this chemical declines in effectiveness, beekeepers find they need to treat more often. I heard that mite-ridden bees constitute a major problem for neighboring beehives. Afterwards I talked about that with ABF president Gene Brandi. He compared it to owning a dog. If your dog had fleas, you wouldn't withhold treatment and just let your dog spread those fleas to other dogs, would you? I know; in the long haul, chemicals aren't the answer. But in the short run, they keep us in the game. When your house is on fire, the short run matters.

At a roundtable we talked about the 2017 EPA directive requiring a veterinary prescription to obtain antibiotics for American foulbrood. I can tell you that confusion reigns, partly because so few vets have any experience with honeybees. Consensus: You might want to consult with a vet before you decide you need antibiotics. The law is the law, but relationships are the grease that make the world work.

I acted brave and talked to strangers. Whenever I got lonely, the Kansas contingent took me in. I'd look around, thinking "Who am I going to go to lunch with?" and Joli and Becky would appear out of thin air, inviting me back to their rental house for pickles and Steve's home-smoked turkey sandwiches.

The best part of all was the people. Beekeepers, salt of the Earth. I felt happy just being among them. You would, too. The ABF, AHPA and CBC are all organizations that promote bees and beekeeping. You ought to join one. I did. Hell, join 'em all! Then you could go to conventions. And if you bump into me at one, maybe we'll go to lunch.

April 2017

The purest food

At the Colorado Beekeepers' meeting in Longmont in December, the conference center provided coffee, tea and honey. The tea was a classy brand called "Tazo." The honey came in little plastic packets. They did not state a country of origin. I could tell you the name of the food distributor on the packet.

I took two packets with my tea and sat down next to Tom. I split one open. How could anything taste so vile? I threw a dollar bill and the other packet on the table in front of Tom and said, "I'll give you a dollar to eat the whole thing."

Tom said, "You can have it back if I don't have to!"

A few minutes later, I noticed him smiling and slipping my dollar bill into his wallet. "You ate that?" I said.

"I put it in my coffee," he said.

I meant for him to eat the honey straight out of the packet to earn his dollar, but still, I had to admire his courage.

Pat from Steamboat Springs rolled into the meeting a little late, after shaking down the slots and rolling the dice in Central City. Down $1,500 and nearly broke at one point, he rallied and walked away a thousand bucks to the good.

That evening at the bar, we talked about honeybee protein supplements. I said, "I buy the bulk bags and mix my own, because it's so hard to get the paper wrappers off the patties."

 Derrick said, "What are you talking about?"

I said, "Well, I take the wrapper off one side of the patties, so the bees can get to it."

Pat said, "You don't have to. That's what those little holes are for. The bees eat right through the paper."

Suddenly a whole table of beekeepers was having a laugh at my expense. But please, I'd rather have egg on my face than keep doing everything backwards for the rest of my life.

A few days later I rode the Aspen Mountain gondola with a skier who runs the movie theater in Steamboat. I asked if he knew Pat. "Sure," he said. "I buy his honey. And we play poker."

Gambling is a mystery to me. I don't have the instincts for it. I used to like to play poker, and the other players at the table liked it when I played, too.

Pat intrigues me. He works fulltime at the supermarket and runs 900 hives on the side -- when he's not running the table. Where does any human being get this kind of energy?

Our keynote speaker was Dr. Malcolm Sanford, retired extension entomologist and professor emeritus at the University of Florida. You might know him as the publisher of the APIS newsletter and a regular contributor to Bee Culture.

In his spare time, Dr. Sanford acts in community plays, dances the tango, and writes prolifically about bees. A former Peace Corps volunteer, he is fluent in Spanish and dabbles in French and Italian. He has a great sense of humor. He reminds me of Pat: How can anyone do so much?

Dr. Sanford spoke at length about the abuse of hive chemicals. He is a strong advocate of burning, rather than treating, American Foulbrood-infected colonies. But how many beekeepers actually do this?

He spoke of the loss of Terramycin as an effective antibiotic and the current use of Tylan as a prophylactic in patties, with the inevitable resistance to Tylan that this will bring. Then there's all that illegal stuff for mites . . .

Beekeepers listening to a lecture on hive chemicals remind me of high school kids getting lectured on underage drinking. They look like little angels sitting there in the auditorium. Then on Saturday night they go out and do what teens do.

Even the manufacturers are looking past their short-term profits to warn of the dangers of chemical abuse. Just look at the effective tools we've lost already: Terramycin, gone. Apistan, gone. Checkmite, gone -- at least for Varroa.

There was a hue and cry from American beekeepers about chloramphenicol in Chinese honey, but what about our own product? Is it what we represent it to be? Are we hypocrites to point fingers?

Remember the Alar scare? It nearly sank the American apple industry 20 years ago. Alar is a plant growth regulator with suspected but unproven links to cancer. In the late 1980s, after the EPA classified Alar as a "probable human carcinogen," CBS 60 minutes ran an expose, and grocery stores nationwide refused to sell Alar-treated fruit. Apple growers found themselves with no market. Adding to the scandal, testing revealed that many growers who pledged to stop applying Alar actually continued to use it.

The public, which had largely regarded apples as a healthy food, came to view them as unsafe to eat.

From the apple growers' economic point of view, it didn't really matter if Alar causes cancer or not. Once the public became alarmed, the damage was done.

Consumers look at honey as a safe, natural food. What could be better for you than honey?

OK, occasionally a customer asks if my honey is "organic." I say, "No. I can't guarantee my bees don't fly onto sprayed fields. How could I?" End of discussion.

More frequently, people inquire if my honey is "raw," although they often have no clear idea what that means.

No one has ever asked me about chemicals in the hive -- not yet.

At an El Salvadoran restaurant the other night, my server put it succinctly: "Honey is a gift from God," she said. "It's the purest food."

April 2010

Eighty-sixed in the almonds

That redheaded Tina got in a little over her head when she volunteered to re-structure the Colorado State Beekeepers Association (CSBA) Master Beekeeper program. Other states run big-budget programs through their universities, but we don't get a nickel from the state of Colorado. Tina quickly learned how challenging it can be to set up a comprehensive beekeeping instruction program on a shoestring. "I'm stressed out and losing weight," she confided.

"Well, take it easy, would you?" I said. "You're no good to me dead."

You have to keep things in perspective. But as president of the venerable Colorado State Beekeepers Association (CSBA), I appreciate Tina's never-flagging efforts. She hangs on like a pit bull. "I never, never give up!" she defiantly proclaims.

CSBA budgeted $1,000 to send Tina to the Western Apiculture Society meeting in Boise in August so she could rub shoulders with a bunch of master beekeeper gurus. When we approved the money, I warned the board that Tina could be cheap. She showed us. She didn't even rent a room! She stayed with local beekeepers, drove, not flew, from Durango in her '94 Volvo wagon. Her total expenditures for her five-day adventure: registration, $175; gas, $246; food, $100; total $521. I am not making this up.

Mere minutes ago, as I sat here banging out his poor epistle, my sweet companion Marilyn came rushing in. "Quick, get your gun!" she cried. The cutest mother skunk got into our hen house and chomped Marilyn's four three-week-old chicks. You could see one tiny set of chick legs back up in the corner where the skunk had trapped herself. "I can't watch," Marilyn said.

Nobody tells me anything. Out of the blue, Marilyn just bought a used food trailer, from which she intends to sell coffee and breakfast pastries. I had no idea, until some woman called wondering when Marilyn was going to pick up her purchase. We went over for a look this morning. Once we get some air in the tires and find the right ball hitch, we can tow it to Marilyn's house in town. The trailer has dainty alpine flowers on it painted by the former owner's Polish wife. It needs some TLC. Marilyn has a business plan, sort of. I'm trying to be a good sport.

Speaking of business plans, mine took a hit when I got cut off from sending my bees to the California almonds. For the past few years I've shipped bees to California in November with Paul's and Derrick's. The bees spend the winter out there and go into the almonds in February. It wouldn't be cost effective for me to hire my own truck for just 100 colonies. In the past, Paul tucked me under his wing and let me share expenses on one of his semi loads. Once the little darlings got dropped off in the Land of Mites and Honey, Derrick took charge, and I got to take the winter off. My colonies returned in March, generally full of bitter almond honey and bearing a nice paycheck. Nothing like kicking off a new season with some financial capital and colonies begging to be split!

But I got sideways with the bee broker in California, a gentleman responsible for leasing thousands of hives to almond growers. He arranges contracts with the growers and tells Derrick where to put bees and how many. So the broker's the kingpin. I'm a bit actor in this huge production. Or was, because the broker recently informed me that my bees are no longer welcome in his operation.

It wasn't the bees' fault. I shipped strong, healthy hives. This had everything to do with bee politics and maybe a touch of lingering bad blood. That's all I'm going to say about that.

I've known the broker for 25 years. Until recently, we got along fine, or at least I thought we did. But now that my little darlings are apis non grata, what's a poor sideliner to do? Even if I found another broker interested in dealing with 100 colonies, I'd still have to somehow get them to California.

They say one door closes, while another opens. There are a thousand ways to make money with bees. I can over-winter colonies here in Colorado, but in March, Colorado bees don't look like their cousins just back from California, bustin' out of their boxes and getting ready to swarm. I've always been able to cover my annual losses with California spring splits.

Seventy might be the new 50, and I suppose it could be, if you retired, took long walks in the park with your sweetheart, rode your bicycle and faithfully attended yoga class. Lifting heavy objects like brood supers is not a recommended senior activity. Maybe at 71 it's time for this beekeeper to cut back and slow down. One serious back injury, and I'd be done. I could sell some bees, and then they'd be somebody else's problem. What's the point of charging full speed into a brick wall? Maybe getting eighty-sixed from the almonds is a wakeup call. The Good Lord works in mysterious ways!

I haven't made a decision, yet. As it stands, my options are wide open.

November 2018

www.ingramcontent.com/pod-product-compliance
Lightning Source LLC
Chambersburg PA
CBHW051658210326
41518CB00021B/2592

9 781912 271887